E. FERNANDEZ CALDAS & DAN H. YAALON (Editors)

VOLCANIC SOILS

WEATHERING AND LANDSCAPE RELATIONSHIPS OF SOILS ON TEPHRA AND BASALT

Selected papers of the "CONGRESO INTERNACIONAL DE SUELOS VOLCANICOS", La Laguna, Tenerife, Canary Islands, Spain, July 1984

CATENA SUPPLEMENT 7

CATENA – A cooperating Journal of the International Society of Soil Science

ISSS - AISS - IBG

CONTENTS

PREFACE

This CATENA SUPPLEMENT contains selected papers presented at the International Meeting on Volcanic Soils held in Tenerife, July 1984. The meeting brought together over 80 scientists from 21 countries, with interest in the origin, nature and properties of soils on tephra and basaltic parent materials and their management. Some 51 invited and contributed papers and 8 posters were presented on a wide range of subjects related to volcanic soils, many of them dealing with weathering and landscape relationships. Classification was also discussed extensively during a six day excursion of the islands of La Palma, Gomera and Lanzarote, which enabled the participants to see the most representative volcanic soils of the Canary Archipelago under a considerable range of climatic regimes and parent material ages.

Because volcanic soils are not a common occurrence in regions where pedology developed and progressed during its early stages, recognition of their specific properties made an impact only in the late forties. The name **Ando** soils, now recognized as a special Great Group in all comprehensive soil classification systems, was coined in 1947 during reconnaissance soil surveys in Japan made by American soil scientists. Subsequently a Meeting on the Classification and Correlation of Soils from Volcanic Ash, sponsored by FAO and UNESCO, was held in Tokyo, Japan, in 1964, in preparation for the Soil Map of the World. This was followed by meetings of a Panel on Volcanic Ash Soils in Latin America, Turrialba, Costa Rica, in 1969 and a second meeting in Pasto, Colombia, in 1972. At the International Conference on Soils with Variable Charge, Palmerston, New Zealand, 1981, the subject of Andosols was discussed intensively. Most recently the definitions of Andepts, as presented in the 1975 U.S. Soil Taxonomy, prompted the establishment of an International Committee on the Classification of Andisols (ICOMAND), chaired by M. Leamy from C.S.I.R., New Zealand, which held a number of international classification workshops, the latest in Chile and Ecuador, in January 1984. The continuous efforts to improve and revise the new classification of these soils is also reflected in some of the papers in this volume.

While Andosols or Andisols formed on tephra (volcanic ash), essentially characterized by low bulk density (less than $0.9 \, g/cm^3$) and a surface complex dominated by active Al, cover worldwide an area of about 100 million hectares (0.8% of the total land area), the vast basaltic plateaus and their associated soils cover worldwide an even greater area, frequently with complex age and landscape relationships. While these soils do not generally belong to the ando group, their pedogenetic pathways are also strongly influenced by the nature and physical properties of the basalt rock. The papers in this volume cannot cover the wide variety of properties of the soils in all these areas, some of which have been reviewed at previous meetings. In this volume there is a certain emphasis on some of the less frequently studied environments and on methods of study and characterization as a means to advance the recognition and classification of these soils.

The Tenerife meeting was sponsored by a number of national and international organizations, including the Autonomous Government of the Canary Islands, the Institute of Ibero American Cooperation in Madrid, the Directorate on Scientific Policy of the Ministry of Education and Science, Madrid, the International Soil Science Society, ORSTOM of France, and ICOMAND. Members and staff of the Department of Soil Science of the University of La Laguna had the actual task of organizing the meeting and the field trips. In editing the book we benefitted from the manuscript reviews by many of our colleagues all over the world, and the capable handling and sponsorship of the CATENA VERLAG. To all those who have extended their help we wish to express warm thanks.

La Laguna and Jerusalem,   E. Fernandez Caldas
Summer 1984   D.H. Yaalon
                                                                              Editors

E. Fernandez Caldas & Dan H. Yaalon (Eds):
VOLCANIC SOILS
CATENA SUPPLEMENT 7, Braunschweig 1985

ESTIMATION OF ALLOPHANE AND HALLOYSITE IN THREE SEQUENCES OF VOLCANIC SOILS, NEW ZEALAND

R.L. **Parfitt** & A.D. **Wilson**, Lower Hutt

ABSTRACT

Allophane was estimated in eight volcanic soils from New Zealand using acid-oxalate and pyrophosphate dissolution. The Al/Si ratio of the allophane was determined and the allophane content of the soil was estimated from the Si extracted in acid-oxalate. Halloysite was estimated by differential thermal analysis of the whole soils.

Allophane predominates in fine grained rhyolitic tephra under high rainfall (2600 mm per year) whereas halloysite predominates under low rainfall (1200 mm). In andesitic tephra under 1200 mm rainfall, imogolite-like allophane is present, but halloysite is present in deeper coarser layers. In rhyolithic alluvium under 1200 mm rainfall halloysite predominates in poorly drained horizons and in coarse textured layers, but allophane predominates in well drained silt loam textured layers.

It appears that Si in soil solution may be controlling the formation of allophane and halloysite; where Si in soil solution is likely to be low, through leaching and free drainage, then allophane predominates; where Si in soil solution is likely to be high, through poor drainage or summer moisture deficit or within larger pumice grains, then halloysite predominates.

1. INTRODUCTION

There is increasing evidence that volcanic ash can weather directly to halloysite as well as to the minerals allophane, imogolite and ferrihydrite. The evidence comes from the following observations:

1.   Halloysite appears to form within glass particles (DIXON & MCKEE 1974).

2.   Halloysite can form in tephra within 800 years (BLEEKER & PARFITT 1974).

3.   Halloysite is thermodynamically more stable than imogolite when Si in soil solution exceeds about 10 μg/l (FARMER et al. 1979).

4.   Halloysite is found in zones of stagnant moisture regime and in silica rich environments (WADA 1977).

There is also evidence that imogolite-like allophane* (Al/Si = 2) forms in large amounts, from volcanic ash, in strongly leached environments, where Si in soil solution is low (PARFITT et al. 1983). Allophane can persist for at least 250 000 years in tephra beds in high rainfall areas in New Zealand (STEVENS & VUCETICH 1985).

On the basis of this evidence, we have suggested the hypothesis that allophane and halloysite formation in volcanic ash soils can be controlled by the Si in soil solution (PARFITT et al. 1983). Imogolite-like allophane (Al/Si = 2) tends to form where Si in soil solution is low and halloysite tends to form where Si in soil solution is high.

---

*   Imogolite-like allophane is used in preference to "proto-imogolite allophane" because the name "proto-imogolite allophane" implies that allophane weathers to imogolite, but there is little evidence for this weathering process.

ISSN 0722–0723 / ISBN 3–923381–06–9

In this paper we have used the method of PARFITT & HENMI (1982) for estimating allophane in clay fractions and have applied this method to whole soils. Halloysite has also been estimated by differential thermal analysis of whole soils. The occurrence of allophane and halloysite is discussed in relationship to climate, parent material and drainage characteristics of soils.

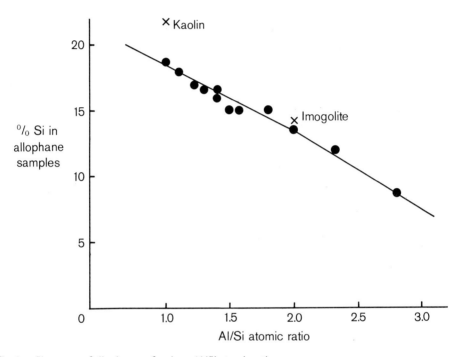

Fig. 1:   Si content of allophanes of various Al/Si atomic ratio.

2. ESTIMATION OF ALLOPHANE IN SOILS

The methods of PARFITT & HENMI (1982) and FARMER et al. (1983) were used to dissolve Al, Fe and Si in non-crystalline weathering products. Both a 4 h extraction with ammonium oxalate (0.2 M pH 3) and a 16 h extraction with sodium pyrophosphate (0.1 M) were used.

Si in acid oxalate extracts comes from the dissolution of allophane, imogolite and ferrihydrite. The ferrihydrite and imogolite content of volcanic ash soils, however, is usually considerably less than the allophane content, so it is possible to use the oxalate Si value to estimate the allophane content of soils.

Fig. 1 gives the Si content of various allophanes according to YOSHINAGA (1966), PARFITT et al. (1980), FARMER et al. (1983) and PARFITT (unpublished). The curve passes through the point for imogolite-like allophane because this is considered to be a reference point in the allophane series. Using Fig. 1 and oxalate Si values it is possible to estimate the allophane content of soils. The Al/Si atomic ratios of soil allophanes can be

Tab. 1: ESTIMATION OF ALLOPHANE USING OXALATE Si IN A) CLAY FRACTION, B) SOIL (FINE EARTH) FROM STRATFORD FINE SANDY LOAM

| (cm) | Dissolution in oxalate | | Dissolution in pyrophosphate | $Al_0 - Al_p/Si_0$ atomic ratio | Allophane content | |
	Al_0 (%)	Si_0 (%)	Al_p (%)		(Si_0) %	(IR) %
A) CLAY FRACTION						
20-30	11.2	5.1	0.2	2.2	40	35
60-70	15.9	9.7	0.1	1.7	65	70
120-130	2.8	3.2	0.04	0.9	15	
B) SOIL (FINE EARTH)						
20-30	2.5	0.94	0.5	2.2	8	
60-70	3.0	1.6	0.2	1.7	11	
120-130	1.4	1.2	0.2	1.0	6	

estimated from $(Al_0-Al_p)/Si_0$ data (Table 1). The Al/Si ratios (Table 1) from clay and soil are in good agreement and indicate that similar materials are dissolved from both soil and clay fractions. The allophane content in the clay fraction was also determined by infrared spectroscopy (Table 1) (PARFITT & HENMI 1982) and there was good agreement with the result obtained from oxalate Si data.

It should be emphasised that this method only gives an estimation of allophane contents and workers should be aware that:

1. Some imogolite and ferrihydrite are also dissolved in acid-oxalate in addition to humus Al, Fe complexes.

2. Possibly acid-oxalate dissolves Al from some chlorite and disordered gibbsites.

3. Pyrophosphate can disperse iron oxides and clay minerals so that pyrophosphate Al, Fe and Si values need to be treated with caution (SCHUPPLI et al. 1983).

3. MATERIALS AND METHODS

3.1. SOILS

The first sequence is a climosequence near Hamilton, where rainfall increases from 1200 mm to 2600 mm per year. The other soil-forming factors are thought to be similar, the soils having formed on the same or similar fine grained tephra beds which are largely rhyolitic (PARFITT et al. 1983). The soils are the Kereone soil, an Andic Dystric Eutrochrept (map grid reference NZMS1 N57/179524); the Ohaupo soil, an Entic Dytrandept (N65/783311) and the Mairoa soil, a Hydric Dystrandept (N82/498782). All the soils are well drained.

The second sequence includes the Stratford soil, an Entic Dystrandept (N129/836369) and the Egmont soil, an Entic Dystrandept (N129/931238), which both formed on andesitic tephra. The Stratford soil is about 10 km closer to the probable tephra source, Mt Egmont, than the Egmont soil which is 40 km from Mt. Egmont. Both soils are well drained.

The third sequence includes three soils of a drainage sequence on a 1 km levee-back-swamp transect, within a relict floodplain near Hamilton. The soils are formed from rhyolitic alluvium with a thin coverbed of Holocene rhyolitic tephra which has probably become mixed with the alluvium. The Waihou silt loam, an Andic Fluventic Dystric Eutrochrept (N57/162619) occurs on an old levee; the Te Puninga silt loam, an Aeric Andaquept (N57/166621) occurs in a mid slope position and the Waitoa silt loam, a Typic Humaquept (N57/173625) occurs in poorly drained backswamp sectors.

3.2. ALLOPHANE AND HALLOYSITE

Allophane in the soil was estimated by the method described above. The Al/Si ratio identified the type of allophane present and the amount of allophane was estimated from the Si extracted in acid oxalate.

Halloysite in the soil was determined from the DTA endotherm at about 530°C. A pure soil halloysite formed from tephra was used as the standard. All samples were lightly ground before DTA.

4. RESULTS AND DISCUSSION

4.1. RHYOLITIC TEPHRA

The Al/Si ratio of the allophane, the allophane contents, the halloysite content of the soils of the climosequence and the bulk density are shown in Table 2.

The water balance has been calculated from the Penman evapotranspiration model assuming a soil water storage of 100 mm (Table 3).

As the rainfall increases from 1200 mm to 2600 mm the allophane content of the soils increases from 1% to 26% and the halloysite content decreases from 35% to 0% (Table 2). The Al/Si ratio of the allophane also increases from about 1 to 2 as the rainfall increases. Halloysite and allophane with Al/Si = 1 have polymerised silicate in their structure (PARFITT et al. 1980), whereas soil allophane with Al/Si = 2 has an imogolite-like structure with orthosilicate. Therefore the degree of polymerisation of silicate in the clay decreases with increasing rainfall.

The Si in soil solution also decreases as both the rainfall increases and the soil drainage flux increases (Table 3). At Maiora, where there is a large through drainage (leaching) of water for about 10 months of the year, the Si in soil solution is low, but at Kereone where there is leaching for about five months in winter and where there is a moisture deficit for about five months in summer, the Si in soil solution is higher (Table 3).

Therefore it appears that when Si in soil solution is high (> about 7 μg/ml) halloysite and allophane (Al/Si = 1) with polymerised silicate are formed. When Si in soil solution is low (< about 7 μg/ml) imogolite-like allophane (Al/Si = 2) with orthosilicate is formed. Thus the amount of Si in soil solution appears to largely control the formation of allophane and halloysite in these soils.

At the Ohaupo site with slightly higher rainfall and leaching and where Si in soil solution is about 7 μg/ml both allophane and halloysite are formed (Table 2). More allophane is present in the upper layers and more halloysite is present in the lower layers. Si is probably leached from the upper layers to the lower layers accounting for the increase in Si content of the allophane with depth as well as the increase in halloysite.

Tab. 2: DISSOLUTION ANALYSIS OF SOILS FORMED FROM SIMILAR RHYOLITIC TEPHRAS UNDER INCREASING RAINFALL; ALLOPHANE AND HALLOYSITE CONTENT, BULK DENSITY OF THE SOILS

		Acid-oxalate			Pyrophosphate		$\frac{Al_0-Al_p}{Si_0}$	Allophane[1]	Halloysite[2]	BD
cm		Al_0 %	Si_0 %	Fe_0 %	Al_p %	Fe_p %	%	%	%	g cm^{-3}
Kereone (1200 mm rainfall)										
A (PC 1042)	0-24	0.6	0.1	0.6	0.2	0.2		<1	30	0.98
Bw	30-41	0.7	0.1	0.6	0.2	0.1		1	30	0.69
2Bw[3]	41-53	0.5	0.4	0.4	0.1	0.1	0.9	2	35	0.82
4Bw[4]	87-101	0.4	0.2	0.3	0.1	0.1	1.2	1	35	1.19
Ohaupo (1400 mm rainfall)										
A (PC 1043)	0-15	3.0	1.4	0.8	0.6	0.2	1.7	10	10	
Bw	36-50	4.1	2.2	1.0	0.3		1.8	15	10	0.59
2Bw[3]	50-60	3.6	2.2	1.0	0.3		1.6	14	15	0.64
4Bw[4]	90-110	1.4	1.0	0.7	0.1		1.3	6	35	0.81
Mairoa (2600 mm rainfal)										
A (PC 1041)	0-18	3.4	0.7	2.8	2.5	2.4	1.4	4	0	0.58
Bw	33-53	8.4	3.5	2.6	1.4	0.4	2.1	26	0	0.41
2Bw[3]	53-60	7.9	3.5	2.0	1.1	0.1	2.0	26	0	0.54
4Bw[4]	81-94	7.1	3.4	3.2	0.8	0.1	1.9	24	0	0.62

[1] from Si_0
[2] from DTA
[3] Aokautere Ash (20,000 y BP)
[4] Rotoehu Ash (42,000 y BP)

Tab. 3: MEAN ANNUAL RAINFALL, DRAINAGE AND DEFICIT AND Si IN SOIL SOLUTION

Soil	Rainfall (mm)	Winter through drainage (mm)	Summer deficit (mm)	Si in soil solution in autumn (μg/ml)
Kereone	1200	400	170	15
Ohaupo	1400	550	140	7
Mairoa	2600	1600	0	2

4.2. ANDESITIC TEPHRA

The allophane and halloysite contents of two soils developed on andesitic tephra (both with mean annual rainfall 1200 mm) are shown in Table 4. The Egmont soil (silt loam) has finer texture than the Stratford soil (fine sandy loam) due to its greater distance from Mt. Egmont. More allophane (19%) has formed in the Bw2 horizon of the Egmont soil than the Stratford soil (10%) presumably because of the smaller particle size in the Egmont soil.

The andesitic tephra has largely weathered to allophane (Al/Si about 2) whereas rhyolitic tephra under the same rainfall (1200 mm, Table 2) has weathered to halloysite. Andesitic tephra has a lower Si content than rhyolitic tephra and presumably the Si in soil solution is low enough to allow allophane (Al/Si = 2) to form.

The Stratford soil, however, contains halloysite in the lower layers and the amounts

Tab. 4: DISSOLUTION ANALYSIS OF SOILS OF DIFFERENT TEXTURE FORMED FROM ANDESITIC TEPHRA (< 20,000 y BP) UNDER 1200 mm RAINFALL

	cm	Acid-oxalate Al_0 %	Si_0 %	Fe_0 %	Pyrophosphate Al_p %	Fe_p %	$\frac{Al_0-Al_p}{Si_0}$	Allophane[1]	Halloysite[2]	BD g cm$^-$
Stratford (fine sandy loam) (PC 1256)										
Bp	0-10	2.1	0.7	0.7	0.7	0.4	1.9	5	0	0.81
Bw1	20-30	2.5	0.9	0.7	0.5	0.2	2.2	8	0	0.88
Bw2	40-50	2.9	1.4	0.8	0.2	0.1	1.9	10	0	0.86
Bw4	60-70	3.0	1.6	0.8	0.2	0.1	1.7	11	0	0.94
C1	100-110	1.4	1.2	1.3	0.1	0.1	1.1	7	4	1.06
C2	120-130	1.4	1.2	1.5	0.2	0.2	1.0	6	14	
Egmont (silt loam) (SB 9557)										
Ap	0-21	3.2	1.1	1.0	0.6	0.1	2.3	9	0	0.75
Bw1	21-55	4.8	2.1	1.4	0.4	0.05	2.1	16	0	0.70
Bw2	55-89	5.0	2.5	1.8	0.3	0.03	2.0	19	0	0.75
2Bw3	89-115	2.9	1.5	1.2	0.2	0.02	1.9	11	4	1.0

[1] From Si_0
[2] From DTA

Tab. 5: DISSOLUTION ANALYSIS OF SOILS FORMED FROM RHYOLITIC VITRIC ALLUVIUM (c. 20,000 y BP) UNDER 1200 mm RAINFALL

	cm	Acid-oxalate Al_0 %	Si_0 %	Fe_0 %	Pyrophosphate Al_p %	Fe_p %	$\frac{Al_0-Al_p}{Si_0}$	Allophane[1] %	Halloysite[2] %
Waitoa (SB 8919)									
Al	0-20	0.53	0.17	0.21	0.25	0.20	1.7	1	10
AB	20-29	0.13	0.03	0.06	0.08	0.08	–	0.2	20
Bg	29-49	0.05	0.02	0.10	0.04	0.06	–	0.1	15
Cg	49-90	0.04	0.02	0.05	0.04	0.04	–	–	13
Te Puninga (SB 8918)									
Al	0-13	2.7	1.1	0.84	0.61	0.37	2.0	8	4
A/B	13-17	4.1	1.7	0.87	0.70	0.37	2.1	13	3
B2	17-39	5.6	2.9	0.71	0.44	0.10	1.9	21	2
Bg	39-57	2.3	1.4	0.45	0.23	0.08	1.6	9	20
Cg	57-85	0.1	0.05	0.1	0.03	0.05	–	–	22
Waihou (SB 8917)									
Al	0-12	2.5	1.1	0.86	0.44	0.15	2.0	8	6
A/B	12-15	1.6	0.70	0.70	0.23	0.10	2.0	5	13
B2	15-22	0.66	0.31	0.96	0.15	0.09	1.7	2	20
B3	22-68	0.26	0.13	0.29	0.07	0.07	1.5	1	15
C1	68-88	1.3	0.83	0.59	0.12	0.06	1.5	5	18

[1] From Si_0
[2] From DTA

increase with depth. The Al/Si ratio of the allophane decreases with depth.

The Stratford soil has a similar moisture regime to the Ohaupo soil (Table 3) and suffi-cient Si probably accumulates in the lower layers for halloysite to form. However, the Ohau-po soil forms from rhyolotic tephra which contains more Si than andesitic tephra. The lower layers of the Stratford soil are sandy loam texture rather than fine sandy loam and it appears that a larger particle size may also favour halloysite formation. Possibly halloysite forms within larger glass and pumice particles (DIXON & MCKEE 1974) where Si concentrations are likely to be high and where, under this moisture regime, Si is not rapidly removed.

4.3. RHYOLITIC ALLUVIUM

The allophane and halloysite contents of soils formed from rhyolotic alluvium under 1200 mm rainfall are shown in Table 5. The sequence consists of:

| Soil | Position | Drainage | Particle Size Distribution | | |
			> 2 mm	2-0.2 mm	0.2-0.02 mm
Waitoa	backswamp	poorly drained	0	15%	40-50%
Te Puninga	mid slope	imperfectly drained	1%	5-10%	40-50%
Waihou	levee	well drained	5%	30-45%	20-30%

The Waitoa soil is poorly drained with an aquic soil water regime and is almost wholly halloysitic. This is consistent with the hypothesis that halloysite forms in a "stagnant mois-ture regime" (WADA 1977) where Si is likely to accumulate in soil solution.

The Te Puninga soil is derived from silt loam over sand alluvium on mid-slope sectors of the levee. In the lower horizons of this soil the predominance of halloysite is attributed to Si accumulation under conditions of poor drainage. In upper horizons above the zone of water table fluctuation, drainage fluxes are probably sufficiently rapid to remove Si, leading to the preferential formation of allophane.

The Waihou soil is formed from coarse textured alluvium on the crest of the levee and is predominantly halloysitic with the halloysite content increasing in lower horizons. 8% allophane occurs in the A horizon and the allophane content generally decreases with depth as does the Al/Si ratio. This soil experiences extended periods of soil water deficit during summer months. Data suggest that drainage is sufficiently rapid to allow allophane and hal-loysite formation in surface horizons. In coarse textured lower horizons, drainage fluxes are likely to be slow due to soil water deficits. This would allow the accumulation of Si in solution and the formation of halloysite. Alternatively, halloysite may be forming in sand size vitric materials below the A horizon as was suggested for the Stratford soil.

The Te Puninga soil (allophanic in upper horizons) and the Kereone soil (halloysitic)

Tab. 6: RAINFALL AND PENMAN POTENTIAL EVAPOTRANSPIRATION FOR HAMILTON

	Jan	Feb	Mar	Apr	May	Jun	Jul	Aug	Sep	Oct	Nov	Dec	Total
Rainfall (mm)	71	81	84	99	112	132	117	117	97	107	91	89	1197
Penman potential evapotranspiration (mm)	125	100	80	45	26	16	20	33	52	80	104	120	801

discussed previously occur in the same region near Hamilton under a similar rainfall (Table 6). The Te Puninga soil, however, has a slightly higher sand content and a larger permeability in upper horizons. This is likely to result in lower Si concentrations in the soil solution and the preferential formation of allophane in the Te Puninga soils.

5. CONCLUSIONS

In fine grained rhyolitic tephra in New Zealand, halloysite tends to form where the annual rainfall is about 1200 mm, where there is a considerable summer moisture deficit and where Si in soil solution is greater than about 7 μg/ml. Allophane is predominant under high rainfall where there is considerable through drainage and where Si in soil solution is low (< 7 μg/ml).

In andesitic tephra under 1200 mm rainfall imogolite-like allophane usually predominates. In coarser tephra, and in deeper layers, halloysite can form together with Si-rich allophane implying high levels of Si are present within pumice particles.

In rhyolitic vitric alluvium, halloysite predominates in poorly drained horizons where Si in soil solution is likely to be high. In well drained horizons, under 1200 mm rainfall, allophane predominates in silt loam textured horizons, but halloysite predominates in sandy loam textured horizons containing larger glass and pumice grains.

REFERENCES

BLEEKER, P. & PARFITT, R.L. (1974): Volcanic ash and its clay mineralogy at Cape Hoskins, New Britain, Papua New Guinea. Geoderma 11, 123-135.

DIXON, J.B. & MCKEE, T.R. (1974): Spherical halloysite formation in a volcanic soil of Mexico. Transactions of the 10th International Conference of Soil Science Moscow VII. 115-124.

FARMER, V.C., SMITH, B.F.L. & TAIT, J.M. (1979): The stability, free energy and heat of formation of imogolite. Clay Minerals 14, 103-107.

FARMER, V.C., RUSSELL, J.D. & SMITH, B.F.L. (1983): Extraction of inorganic forms of translocated Al, Fe and Si from a podzol Bs horizon. Journal of Soil Science 34, 571-576.

PARFITT, R.L., FURKERT, R.J. & HENMI, T. (1980): Identification and structure of two types of of allophane from volcanic ash soils and tephra. Clays and Clay Minerals 28, 328-334.

PARFITT, R.L. & HENMI, T. (1982): Comparison of an oxalate-extraction method and an infrared spectroscopic method for determining allophane in soil clays. Soil Science and Plant Nutrition 28, 183-190.

PARFITT, R.L., RUSSEL, M. & ORBELL, G.E. (1983): Weathering sequence of soils from volcanic ash involving allophane and halloysite, New Zealand. Geoderma 29, 41-57.

SCHUPPLI, P.A., ROSS, G.J. & MCKEAGUE, J.A. (1983): The effective removal of suspended materials from pyrophosphate extracts of soils from tropical and temperate regions. Soil Science Society of America Journal 47, 1026-1032.

STEVENS, K.F. & VUCETICH, G.C. (1985): Weathering of Upper Quaternary tephras in New Zealand. Part II Clay Minerals. Chemical Geology (in press).

WADA, K. (1977): Allophane and imogolite. In: J.B. Dixon and S.B. Weed (Editors): Minerals in Soil Environments. Soil Science Society of America Madison. 603-638.

YOSHINAGA, N. (1966): Chemical composition and some thermal data of eighteen allophanes from Ando soils and weathered pumices. Soil Science and Plant Nutrition 12, 47-54.

Address of authors:
R.L. Parfitt and A.D. Wilson, Soil Bureau, DSIR,
Lower Hutt, New Zealand

E. Fernandez Caldas & Dan H. Yaalon (Eds):
VOLCANIC SOILS
CATENA SUPPLEMENT 7, Braunschweig 1985

STUDY OF AMMONIUM OXALATE REACTIVITY AT pH 6.3 (Ro) IN DIFFERENT TYPES OF SOILS WITH VARIABLE CHARGE. I

J.M. Hernández Moreno, V. Cubas García, A. González Batista &
E. Fernández Caldas, La Laguna

ABSTRACT

The oxalate reactivity at pH 6.3, Ro, which makes use of the ligand exchange between oxalate and OH reactive groups of allophane, allophane-like materials and organo-mineral complexes has been studied in different types of andic soils and other soils with variable charge. The Ro values are very sensitive to the pre-drying of the samples and they have been determined on moist, air-dried, P_2O_5-dried and 105°C heated samples. Generally, there is a marked increase of the Ro values upon drying up with P_2O_5. Heating at 105°C reduces Ro values in the Al-rich allophanes while in the Si-rich, Ro values increase. This behaviour has been explained in terms of the reduction of surface area and increase in surface acidity upon drying and heating the allophane clays. A fair correlation has been found between Ro values and Al extracted with oxalate pH 3 except in the Al-rich saples in which the ratio Ro/Al_3 is higher. As would be expected, high Ro values are associated with high anion exchange capacity values. Ro thus gives a great deal of information about the physicochemical behaviour of allophanic soils.

1. INTRODUCTION

The characterization of the properties of Andisols and other soils of variable charge has been based to a great extent on the quantification of materials by selective dissolution: allophanic compounds, amorphous oxihydroxides or organo-mineral complexes. Recently it was considered more convenient to define these materials in terms of their activity, especially active-Al, rather than their quantity (WADA 1980, PARFITT 1980). With this in mind we have studied a method based on the exchange kinetics of OH ions by the oxalate anion (oxalate reactivity).

The results obtained with allophanic clays (GONZALEZ BATISTA et al. 1982) and the simplicity and quickness of the method prompted us to apply this technique to soils. In the present work we study the oxalate reactivity in soils of variable charge containing compounds of high surface activity (allophane, imogolite, amorphous Al and Fe). These surfaces are very sensitive to drying and heating and in this work the changes undergone are studied through oxalate reactivity. In this first work the results obtained on a great number of soils from different places, particularly Andisols, Andic soils and some Podsols, are presented.

2. SOILS

The soils studied belong to several types: Andisols, Andic soils intergraded to ferralitic-type soils and some Podsols. (The classification of these soils is given later in Table 3).

ISSN 0722–0723 / ISBN 3–923381–06–9

   – Soils of the Canary Islands. These soils are described in the "Guia de Campo" of the International Congress of Volcanic Soils (Tenerife, 1984).
   – Soils of Chile and Ecuador. These are described in the book on description of profiles from the VI International Soil Workshop (Chile–Ecuador, 1984).
   – Soils of Japan. These are described in the publication on "International Correlation of Kuroboku and Related Soils" (WADA, 1983).
   – The Akaka soil (Hawaii) has been studied, among other authors, by EL SWAIFY & SAYGEH (1975) and BALASUBRAMANI.\N & KANEHIRO (1978).
   – The Komagadake soil was studied by NAKAI & YOSHINAGA (1980). Soil nr. 1419 (Galicia, Spain) was studied by MACIAS VAZQUEZ (personal communication).
   The most important properties of these soils have been described in the above-mentioned publications.

3. AMMONIUM OXALATE KINETICS (Fraction < 2 mm)

   Samples of soils (< 2 mm) in different conditions of moistness were used: field moist, air-dried, dried over P_2O_5 and heated at 105° (24 h).

3.1. WITH AUTOMATIC TITRATOR

   The system employed consisted of a centrifuge tube where 400 mg of soil (fraction < 2 mm) and 20 ml of saturated ammonium oxalate solution in the presence of 0.1 g of solid ammonium oxalate were placed. The pH was maintained at a value of 6.3 (or the value being studied) for 25 minutes by means of an automatic titrator (GDA), with constant stirring (magnetic stirrer). The micro electrode used was an INGOLD-2233. The values presented in this work were obtained at 20°C ($\pm$ 1°C).
   The number of OH groups displaced during the reaction is given by the number of HCl millimoles (1 N or 0.1 N) consumed during the 25 minutes. We call this amount oxalate reactivity, Ro.
   As it will be seen later, oxalate reactivity depends on the moisture content of the samples, and the symbol Ro will refer to samples dried over P_2O_5.

3.2. NORMAL TITRATION

   The determination of Ro was carried out by normal titration with results similar to those above. Using 2 g of soil and 100 ml of oxalate solution and titrating with 0.2 N HCl, the value (ml) of HCl consumed multiplied by 10 gives directly the value of Ro. In these conditions, the most frequent interval is from 0 to 10 ml.

3.3. ANALYSIS OF THE SUPERNATANT SOLUTION FROM REACTION AT pH 6.3

   When the reaction ended after 25 minutes, the suspensions were centrifuged immediately. Al, Fe. and Si were analysed in the supernatant by Atomic Absorption (Perkin-Elmer-603).

3.4. CATION AND ANION EXCHANGE CAPACITIES (CEC and AEC)

The method employed was according to GILLMAN (1979) ("Compulsive Exchange).

4. RESULTS

We found that most important conditions to control when carrying out the oxalate reactivity were: pH, temperature, excess of solid oxalate and water content of the samples.

Tab. 1:   Ro DEPENDENCE OF pH IN AIR DRIED SAMPLES (< 0.2 mm).

Soils		Ro	
		pH = 6,3	pH = 6
GUINATE	Bw/C	20	33
LA ESPERANZA	Bw/C_1	61	102
LAS RAICES	2B	13	21
CHIO	C_1	28	53

4.1. pH

The values of Ro are very sensitive to the pH maintained during the reaction (Table 1).
In this work we maintained the value of pH 6.3, previously used (GONZALEZ BATISTA et al. 1982) for the following reasons:
– the buffer capacity of the oxalate solution is suficiently low to allow the rapid release of OH^- groups.
– at values of pH below 6.3, reactivity may be very high so that the dissolution is substantial; our objective was to achieve a "surface" reaction only.
At pH values above, 6.3, some soils, such as the Vitrandepts, give Ro values which are too low to be accurately determined.
Thus, at pH 6.3, values of Ro in little reactive and very reactive soils can be compared within the same soil type.
With regard to the influence of the pH of the soil on the Ro values, two cases should be differentiated. In the case of acid soils, it might be thought that Ro values were underestimated, i.e. that there is a consumption of OH ions by the soil changing from actual pH to 6.3.
The buffer capacity of the oxalate solution, however, increases rapidly below pH 6 and in the soils studied it was seen that very low values of initial soil pH do not substantially influence the Ro values. In the case of soils having pH above 6.3, pH values above this value were only found in some Vitrandepts (pH at an ionic strength comparable to that of the oxalate solution, approximately 1N). The error – in excess in these cases – is small (<6 meq/100 g) and maintains these soils within their reactivity range. When calcium carbonate is present in the soil, the situation is more complex because of the consumption of protons. This case is not studied in the present work.

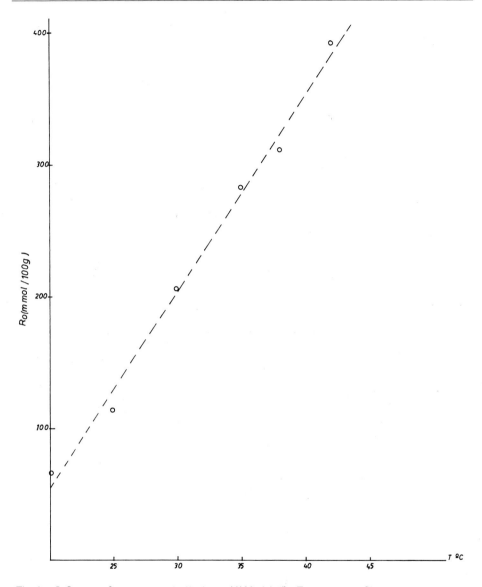

Fig. 1:   Influence of temperature in Ro (mmol/100 g) in La Esperanza profile.

4.2.   TEMPERATURE

The temperature notably influences the Ro values. The relationships observed between Ro and T (example in fig. 1) seem to indicate that the increase of Ro with the temperature can be attributed in part to the increase in solubility of the ammonium oxalate with the temperature.

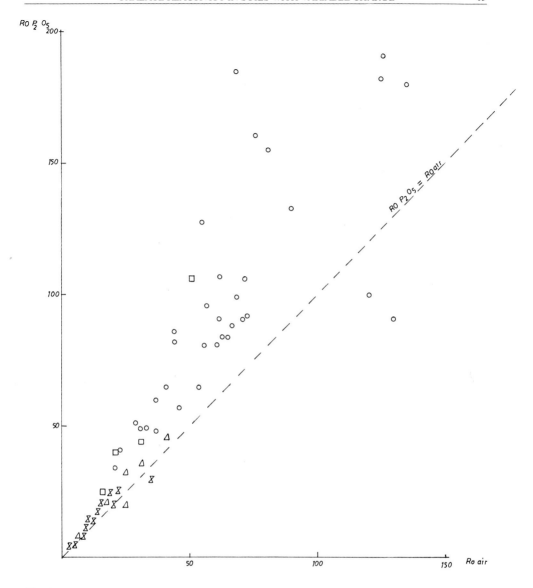

Fig. 2: Ro (mmol/100 g) in P_2O_5-dried and air-dried samples. ○ Andic Andisols; 𝕏 Vitric Andisols; △ Andisols Intergrades; □ Podsols.

4.3. EXCESS OF SOLID AMMONIUM OXALATE DURING THE REACTION

The excess of solid reagent is essential for the oxalate concentration to be maintained constant, (especially in the case of very reactive samples), and to ensure, in the case of very acid soils, that the buffer capacity is maintained.

4.4. INFLUENCE OF PREVIOUS DRYING AND HEATING OF THE SAMPLES

The effect of drying leads to an increase of reactivity which attains the maximum values in the samples dried over P_2O_5 (of the 67 samples studied we have only found two significant exceptions to this tendency). With a view to establishing reproducible reference conditions for the water content of the samples, the oxalate reactivity in samples dried over P_2O_5 has been fundamental in this work. However, the correlation obtained between the reactivity values in P_2O_5-samples and air-dried samples allowed the values obtained in these last conditions (fig. 2) to be used on many occasions.

With a view to standardising the nomenclature, Ro will refer always to the reactivity values of the samples dried over P_2O_5, unless otherwise stated in a footnote. Ro_H , Ro_S and Ro_{105} indicated, respectively, the reacivity values in the moist sample, air-dried and heated to 105°C. In the same manner, the difference between the reactivity values determined in the P_2O_5 sample and the 105°C sample has been called $\triangle Ro$. $\triangle Ro = Ro-Ro_{105}$, always taking Ro as minuend. The different reactivity values and $\triangle Ro$ of the soils studied are presented in Table 3. It was observed that, contrary to the effect of drying, the effect of heating on Ro was variable, i.e. both positive and negative Ro values were obtained.

5. DISCUSSION

In previous work, it was observed that the oxalate reactivity values at pH 6.3 for allophanic clays was highest in those materials with high Al/Si ratios, associated with high ZPC and anion exchange capacity values (GONZALEZ BATISTA et al. 1982).

When applying the technique to the whole soil (fraction < 2 mm) it must be taken into account that the Ro values are not only going to depend on the type of allophane or active material present, but also on their ammount. Considering that at the present time extraction with ammonium oxalate pH 3 is recommended for the quantitative estimation of the allophanic materials (PARFITT 1980, RUSSELL et al. 1981), the Ro values have been compared with those corresponding to the %Al, extracted with oxalate pH3 (fig. 3). It is observed that very different values of Ro may correspond to any one value of %Al, indicating that Ro not only depends on the amount of extractable material but also on other characteristics.

The relative amount of the elements (Al, Fe, Si) dissolved during the oxalate reaction at pH 6.3 can afford important information on the type or "parts" of the material they dissolve. The Ro and (Al + Fe/Si) are related in fig. 4. It is observed that, in general, the A horizons and/or those associated with high Al pyrophosphate values give the highest Al/Si relationships (maximum value found: 17). At the other extreme, the Vitrandepts give the lowest values (2-1). These results suggest that in the oxalate reaction pH 6.3, Al-OH and (Fe-OH) groups mainly are involved. These ratios seem high even in the case of the Vitrandepts, if it is borne in mind that the Al/Si ratios in their <2 mm fractions may be around 0.5 or less.

The anion and cation exchange capacity values of some soils are given in Table 2. Although no defined relationship can be observed, high Ro values are generally associated with samples with a high anion exchange capacity.

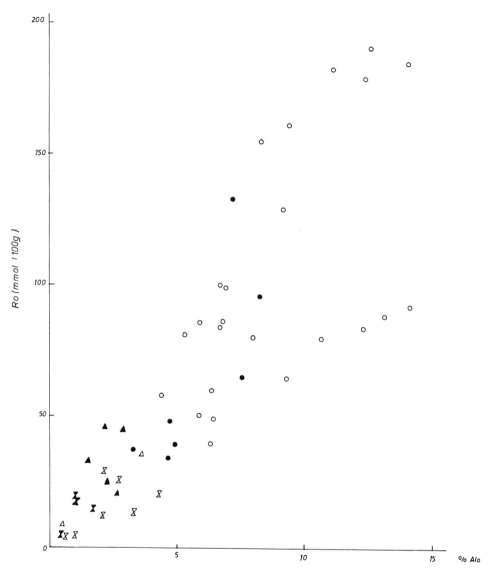

Fig. 3: Ro (mmol/100 g) and %Al extracted in ammonium oxalate pH 3. (for symbols see fig. 2). Black symbols represent A horizons.

5.1. INFLUENCE OF DRYING AND HEATING ON Ro VALUES

The great sensitivity of allophanic materials to drying and heating is well known and is manifested by important variations in their physical and chemical properties. Surface area, porosity, ionic exchange characteristics and surface acidity should be stressed particularly.

The work carried out in this regard allows, in a first approach, an attempt to be made to explain the variations observed in the samples studied (Table 3), while results obtained using

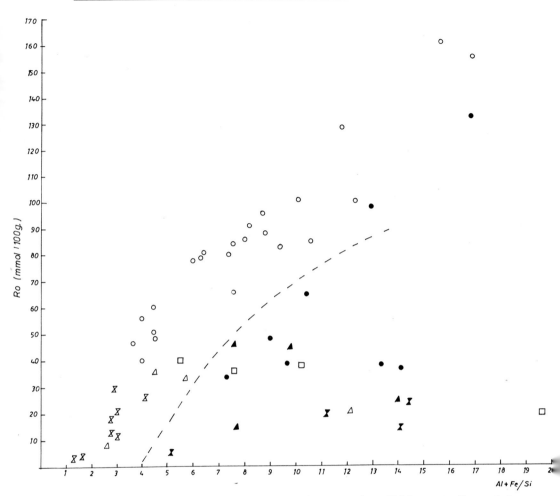

Fig. 4:   Ro (mmol/100 g) and (Al + Fe)/Si (molar) in the oxalate pH 6.3 extracts. (for symbols see fig. 2). Black symbols represent A horizons.

Tab. 2:   ANION AND CATION EXCHANGE CAPACITIES AEC & CEC AND Ro OF AIR DRIED SAMPLES.

Soils		AEC	CEC	Ro
LA ESPERANZA	A_1	4,6	14,2	37
	Bw/C_1	14,0	12,0	56
	Bw/C_2	12,0	8,5	63
BARLOVENTO	Bw_1	8,8	9,0	81
	Bw_5	10,5	4,5	55
BONA	Bw_1	9,8	13,0	127
	Bw_2	11,6	12,1	135
LAS BREÑAS	A_1	2,2	9,5	21
	$2C_2$	2,5	19,0	16
	$4C$	1,5	14,3	13
CHIO	A_1	1,1	18,5	12

Tab. 3: Ro VALUES AT DIFFERENT MOISTURE CONTENTS, Al/Si RATIOS IN OXALATE pH 6.3 AND OXALATE pH 3 EXTRACTS. Al OXALATE pH 3, Al PYROPHOSPHATE AND MOISTURE CONTENT IN THE P_2O_5 DRIED SAMPLES.

Soils		Ro_H	Ro_S	Ro	Ro_{105}	ΔRo_{P-S}	ΔRo	$(Al/Si)_{P_2O_5}$	$\%Al_0$	$\%Al\ P$	$\%H_2O\ P_2O_5$	$(Al/Si)_0$
BARLOVENTO Hydrandepts	Ap_2	45	90	133	95	43	38	15,0	7,2	2,9	7,5	4,4
	Bw_1	71	81	155	137	74	18	15,8	8,3	3,3	6,8	4,5
	Bw_2	39	76	161	123	85	38	14,3	9,4	2,6	7,8	4,2
	Bw_3	26	44	86	80	42	6	10,2	5,9	1,4	5,2	3,8
	Bw_4	49	68	185	94	117	91	13,2	14,2	1,8	8,5	3,3
	Bw_5	37	55	128	78	73	50	11,5	9,3	1,2	–	3,4
	Bw_7	27	44	82	60	38	22	9,1	5,3	1,1	5,3	3,4
BONA Hydrandepts	A_1	56	57	96	31	39	65	8,3	8,3	3,1	8,5	4,1
	Bw_1	73	127	191	17	64	174	12,8	12,5	2,6	9,8	3,3
	Bw_2	58	135	179	16	44	163	12,2	12,4	2,2	11,2	3,3
	Bw_3	56	125	183	33	58	150	11,6	11,1	1,8	11,3	2,9
	Bw_4	34	96	99	55	30	44	9,5	6,9	1,8	7,4	3,6
GARAFIA Dystrandepts	A_1	48	54	65	21	11	44	10,0	7,6	1,5	7,5	2,8
	Bw_1	68	67	88	20	21	68	8,5	13,1	1,3	10,2	2,7
	Bw_2	58	73	92	28	19	64	8,0	14,1	1,1	10,4	2,6
	Bw_3	49	65	84	25	19	59	6,1	12,3	0,7	9,3	2,5
AGUAMANSA Dystrandepts	A_1	–	33	39	35	6	4	9,0	4,9	1,0	4,7	2,3
	Bw_1	–	29	51	29	12	22	4,1	5,8	0,4	5,0	1,9
	Bw_2	–	31	49	36	18	13	4,2	6,4	0,4	4,9	1,8
	Bw_3	–	37	60	52	23	8	4,1	6,3	0,4	5,0	1,8
	$2Bw$	–	61	81	42	20	33	7,1	10,6	0,5	6,0	1,9
LA ESPERANZA Dystrandepts	A_1	–	37	48	43	11	5	8,5	4,6	0,6	4,4	2,8
	Bw/C_1	–	56	81	64	25	17	5,9	7,9	0,4	6,2	2,3
	Bw/C_2	–	63	84	65	21	19	6,8	6,7	0,4	7,2	2,2

Tab. 3:   Continued

CHILE–1 Typic Dystrandepts											
2Bs$_1$	—	46	57	56	11	1	3,6	4,4	—	5,5	2,3
CHILE–7 Typic Dystrandepts											
BA	—	62	107	45	45	62	—	4,0	1,5	7,8	2,6
Bs$_1$	—	72	91	42	19	49	—	4,1	1,2	5,3	2,4
Bcs	—	72	106	43	34	63	—	5,8	0,6	5,2	2,0
CHILE–9 Aquic Dystrandepts											
E	—	110	100	55	–10	45	11,9	6,7	1,4	9,5	2,6
Bs$_1$	67	130	86	42	–44	44	7,8	6,8	1,3	8,4	2,5
CHIO Vitrandepts											
A$_1$	—	12	15	20	3	–5	12,0	1,7	0,40	3,2	2,5
C$_1$	—	10	13	18	3	–5	2,1	2,1	0,11	3,4	1,3
C$_3$	—	5	5	11	0	–6	1,2	1,0	0,05	2,6	1,5
C$_4$	—	4	5	8	1	–3	1,0	0,6	0,04	1,5	1,5
ECUADOR–1 Vitrandepts											
A$_1$	—	7	6	12	–1	–6	4,1	0,5	—	1,5	1,7
ECUADOR–5 Vitrandepts											
3Ab	—	23	27	27	4	0	2,8	2,7	—	4,4	1,6
3C$_2$	19	36	30	37	–6	–7	3,0	2,1	—	3,6	1,4
ECUADOR–6 Dystrandepts											
A$_2$	19	20	20	28	0	–8	2,0	1,0	—	2,4	2,0
GUINATE Vitrandepts Camborthids											
Bw/C	—	14	18	27	4	–9	2,2	1,1	0,10	0,5	2,0
LAS LAJAS Palehumults											
A$_1$	—	41	46	55	5	–9	6,6	2,2	2,30	2,0	7,3
Bw	—	25	33	45	8	–12	4,4	1,5	2,20	2,2	7,5
LAS HAYAS Haplumbrepts											
A	—	—	45	54	—	–9	8,9	2,9	2,10	7,0	4,8
JAPAN I–3 Andic Haplumbrepts											
A$_{11}$	28	—	38	29	—	9	13,2	—	—	4,0	—
IIIB$_2$	12	—	20	18	—	2	11,0	—	—	3,2	—

Soil	Horizon											
JAPAN 1-4 Hydric Dystrandepts	VB₂	29	–	47	32	–	15	3,0	–	–	4,5	–
JAPAN T-1 Hydric Dystrandepts	IIB₂	37	–	77	47	–	30	5,5	–	–	6,1	–
JAPAN T-2 Hydric Dystrandepts	A	67	–	89	51	–	38	12,6	–	–	5,1	–
LAS BREÑAS Dystrandepts	A₁	18	21	34	25	13	9	6,4	4,7	0,6	5,7	2,0
	C₁	–	23	40	33	17	7	3,4	6,3	0,3	5,3	1,4
	2C₂	10	16	21	40	5	– 19	2,4	4,3	0,2	2,5	1,5
	4C	10	13	14	35	1	21	2,2	3,3	0,2	3,4	1,5
GARAJONAY Haplumbrepts	A	–	19	25	21	6	4	2,3	2,30	1,50	2,3	5,7
LAS RAICES Dystrandepts	A₁	–	18	21	24	3	– 3	7,1	2,70	0,50	3,0	2,8
	Bw	–	31	36	37	5	– 1	4,2	3,60	0,20	3,6	2,3
	2B	–	7	9	13	2	– 4	2,0	0,50	0,60	2,1	2,5
AKAKA (64-114 cm) Hydrandepts		22	–	60	24	–	36	15,6	–	–	3,0	–
KOMAGADAKE Podsol–Japan	A₂	–	0	0	0	–	–	9,0	–	–	0,6	–
	B₂₁	–	31	44	30	13	14	17,0	–	–	3,2	–
	B₂₂	–	51	107	65	56	42	16,0	–	–	8,3	–
GALICIA Podsol (1419)	Bh	16	–	25	29	–	– 4	12,5	0,75	1,78	0,8	–
	Bs	22	–	41	50	–	– 9	4,6	1,23	0,79	0,8	–
CHILE–6 Podsolic Soil?	A₁	–	26	22	13	– 4	9	15,4	0,80	–	2,8	4,1
	A₂	–	49	38	33	– 11	5	8,3	0,20	–	2,1	4,1
	C₁	47	66	36	23	– 30	13	6,5	0,90	–	2,0	2,3

Symbol explanation: $(Al/Si)_{P_2O_5}$: In Ro extract; %Alo: Oxalate pH 3; %Alp: Pyrophosphate; %H_2O P_2O_5 : %H_2O in P_2O_5-dried sample; $(Al/Si)_0$: Oxalate pH 3 "pure" reference models become available.

The results which seem most interesting are those relating to the change in reactivity undergone by the samples from drying with P_2O_5 to heating at 105°C, $\triangle Ro$. Soils giving positive $\triangle Ro$ values, i.e., Ro diminishes on heating to 105°C, have one or more of the following characteristics: they are associated with high Ro values, the material extracted with ammonium oxalate pH 3 has high Al/Si ratios and they posses high anion exchange capacity and ZPC values.

The concretions of gibbsite from the Barlovento profile, of low crystallinity, also give positive values of $\triangle Ro$.

The soils which give negative values of $\triangle Ro$, i.e. reactivity increases on heating to 105°C have one or more of the following characteristics: they are associated with low Ro values, the material extracted with ammonium oxalate (pH 3) have low Al/Si ratios, they contain important amounts of kaolinite-type clay minerals. and the values of anion exchange capacity are low. This behaviour was observed in vitric Andisols, Andisols Intergrades to Ferralitic Soils and in some Bh and Bs horizons of Podsols.

The reasons for the variations in Ro on drying and heating must be sought in the modifications of the surface area and surface chemical characteristics.

5.2. SURFACE AREA

The surface area and the porosity of the allophanic materials decrease upon drying the sample and particularly upon heating it (EGASHIRA & AOMINE 1974, ROUSSEAUX & WARKENTIN 1976). The results obtained by these authors seem to indicate that the most Al-rich materials are the most affected by heating, on the other hand, the reversibility of the process must also be taken into account.

In this regard, UEHARA & GILLMAN (1981) indicate that the degree of reversibility in the amorphous materials may depend on the Al/Si ratio, the Al-rich materials showing the highest irreversibility.

5.3. SURFACE CHEMICAL CHARACTERISTICS

The increase of Ro values on drying seems to indicate that there is an increase in surface activity. Indeed many authors have found that chemical surface properties of allophanic materials change with their water content. HENMI & WADA (1974) and HENMI (1977) have shown that the surface acidity of allophane and imogolite increases upon drying and heating and that this effect is greater in the siliceous allophanes (Al/Si ~ 1) than in the Al-ich allophanes and imologite. BALASUBRAMANIAN & KANEHIRO (1978) found that in Hydrandepts the ZPC values diminish on drying and heating and attribute this phenomenon to the fact that the new surfaces produced during the treatment display greater acidity. HARADA & WADA (1974) found that the cation exchange capacity of allophanic soils increases on drying and heating.

On the other hand, it is known that in the anion adsorption on oxide surfaces by ligand exchange, the anions of weak acids display a maximum adsorption at pH values near their pK; the adsorption is made possible by the presence of protons on the oxide surface or derived from the dissociation of the weak acid (HINGSTON et al. 1972).

All these results seem to indicate that the increase of Ro on drying is favoured by an increase of acidity at the hydroxilated surfaces.

At the stage of this study we can think that there are probably two phenomena which act in opposite direction on drying and heating:

1) The surface area or porosity decrease.
2) The acidity of hydroxilated surface increases.

With the process of drying the second effect would seem to predominate and Ro values increase in all the samples.

On heating, the reduction of surface area or porosity may be important and little revesible possibly leading to a reduction of Ro.

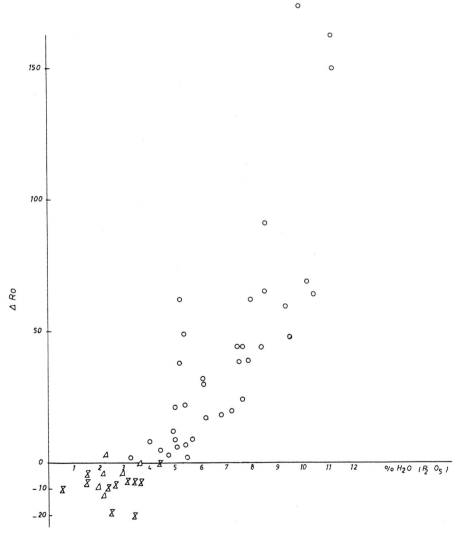

Fig. 5:   Ro (mmol/100 g) and %H₂O content in P_2O_5-dried samples (105°C basis). (for symbols see fig. 2).

In relation to the above hypothesis it is interesting to observe fig. 5 where $\triangle Ro$ is represented against the water content over P_2O_5 on 105°C basis. The higher water content in Al-rich samples is an agreement with the findings of ROSSEAUX & WARKENTIN (1976) that allophanes with a high Al content show a high micropore content and consequently higher water retention. A collapsation of the pores upon heating at 105°C would lead to a more important reduction of available surface in the case of Al-rich samples and this can probably explain the higher decrease of the Ro values in these samples.

With regard to the clay minerals, an increase of Ro on heating can also be explained since the surface area is not substantially modified and an increase of surface acidity takes place (HENMI & WADA 1974).

In the case of the organo-mineral compounds the situation is more complex because of the great variety of the situations which can be found. On principle, an increase of acidity can be expected as the water content decreases (MORTLAND 1970). The humic substances may block, to a lesser or greater degree, the OH groups linked to Al and Fe (PARFITT 1978). It can be observed in fig. 4 how the samples corresponding to A horizons or with a relatively high content of Al and Fe extractable in pyrophosphate are situated in the low zone, to the right, i.e., for a given reactivity value, the dissolution produced with oxalate pH 6.3 extracts much greater amount of Al and Fe.

The results obtained indicate that Ro parameters may give a valuable information on the surface OH groups activity of allophane materials and amorphous oxihydroxides as well as on Fe and Al organo-mineral compounds.

Some important physical and chemical properties of soils of variable charge rich in amorphous materials are manifested through these parameters. We think it is of great interest to test these parameters in a greater number of soils and reference models samples.

REFERENCES

BALASUBRAMANIAN, V. & KANEHIRO, Y. (1978): Surface chemistry of the Hydrandepts and its relations to nitrate adsorption as affected by profile depth and dehydration. J. Soil Sci. **29**, 47-57.

EGASHIRA, K. & AOMINE, S. (1974): Effects of drying and heating on the surface area of allophane and imogolite. Clay Sci. **4**, 231-242.

El SWAIFY, S.A. & SAYEGH, A.H. (1975): Charge characteristics of an oxisol and of Inceptisol from Hawaii. Soil Sci. **120**, 49-56.

GILLMAN, G. (1979): A proposed method for the measurement of exchange properties of highly weathered soils. Aust. J. Soils Res. **17**, 129-139.

GONZALEZ BATISTA, A., GARCIA, J., HERNANDEZ MORENO, J.M. & FERNANDEZ CALDAS, E. (1982): Estudio de la cinética de la reacción del oxalato amónico con arcillas alofánicas. Anales de Edafología y Agrobiología. T. XLI. núms. **5-6**, 915-926.

HARADA, Y. & WADA, K. (1974): Effects of previous drying on the measured cation- and anion-exchange capacities of andosoils. transactions of the 10th International Congress of Soil Science. Vol. **II**, 248-256. Moscow.

HENMI, T. (1977): The dependence of surface acidity on chemical composition of allophane. Clay Min. **12**, 356-368.

HENMI, T. & WADA, K. (1975): Surface acidity of imogolite and allophane. Clay Min. **10**, 231-245.

HINGSTON, F.J., POSNER, A.M. & QUIRK, J.P. (1972): Anion adsorption by Goethite and Gibbsite. I. The role of the proton in determining adsorption envelopes. J. Soil Science, **23**, 2, 177-192.

VI INTERNATIONAL SOIL CLASSIFICATION WORKSHOP. CHILE–ECUADOR, 1984.

MACIAS VAZQUEZ, F. – Departamento de Edafología. Facultad de Biología. Santiago de Compostela.

MORTLAND, M.M. (1970): Clay organic complexes and interactions. Adv. Agron. **22**, 73-117.

NAKAI, M. & YOSHINAGA, N. (1980): Fibrous goethite in some soils from Japan and Scotland. Geoderma, **24**, 143-158.

PARFITT, R.L. (1978): Anion adsorption by soils and soil materials. Advances in Agronomy. Vol. **30**, 1-50.

PARFITT, R.L. (1980): Chemical properties of variable charge soils. In: Soils with variable charge. B.K.G. THENG (Ed.). N.Z., 448.

ROUSSEAUX, J.M. & WARKENTIN, B.P. (1976): Surface properties and forces holding water in allophane soils. Soil Sci. Am. J. Vol. **40**, 446-451.

RUSSELL, M., PARFITT, R.L. & CLARIDGE, G.G. (1981): Estimation of the amount of allophane and other materials in the clay fraction of Egmont Loam profile and other volcanic ash soils, N.Z. Aust. J. Soil. Res. **19**, 185-195.

UEHARA, G. & GILLMAN, G. (1981): The mineralogy, chemistry and physics of tropical soils with variable charge clays. W. Tropical Agr. Series. no. **4**, 170.

WADA, K. (1980): Mineralogical characteristics of Andisols. In: Soils with variable charge. B.K.G. THENG (Ed.) N.Z., 448.

WADA, K. (1983): International correlation of Kuroboku and related soils. Grant-in-Aid for Cooperative Research No. 563006 (1981-1982).

Address of authors:
J.M. Hernández Moreno, V. Cubas García, A. González Batista and E. Fernández Caldas
Departamento de Edafología, Facultad de Biologia. Universidad de La Laguna Carretera Vieja de La Esperanza.
La Laguna, Tenerife. (Canary Islands). Spain

E. Fernandez Caldas & Dan H. Yaalon (Eds):
VOLCANIC SOILS
CATENA SUPPLEMENT 7, Braunschweig 1985

BEHAVIOUR OF OXALATE REACTIVITY (Ro) IN DIFFERENT TYPES OF ANDISOLS. II

E. Fernández Caldas, J. Hernández Moreno, M.L. Tejedor Salguero,
A. Gonzalez Batista & V. Cubas Garcia, La Laguna

ABSTRACT

In this work, an attempt is made to apply Ro as a parameter to characterise Andisols, since Ro integrates many important properties of allophanic materials.

Volcanic soils from the Canary Islands, Japan, Chile and Ecuador, belonging to Vitric, Andic and Andisol Intergrades have been selected for study. The behaviour of Ro in each group has been studied relating it with diagnostic criteria such as %Al (oxalate pH 3), %P retention, bulk density and water retention. A discussion is also made in relation to the classification of these soils in different systems. The ability of amorphous compounds to undergo important changes upon heating and drying has also been studied through Ro in dried and heated samples, using $\triangle$Ro (difference of Ro values in P_2O_5 dried samples and 105°C heated samples).

In the soils studied, Ro and $\triangle$Ro values allowed us, once the Order was defined, to establish a continuum between Andic and Vitric Andisols as members of a unique family, irrespective of Vitric and Andic characteristics.

These and other observed relations show a wide span of possibilities to be explored in order to get quick and reliable information on the characteristics of Andisols.

1. INTRODUCTION

In previous work (GONZALEZ BATISTA et al. 1982) and in the preceding paper (HERNANDEZ MORENO et al. 1985), the oxalate reactivity at pH 6.3, Ro, in the fraction $< 2\,\mu m$ and whole soil of Andisols and other soils of variable charge has been studied (Ro = number of mmol/100 g of OH groups released during the reaction of ammonium oxalate with samples, controled at pH 6.3). The Ro parameter is especially related to the active aluminium and permits allophanic materials of different characteristics to be differentiated. In this regard it should be noted that allophanic siliceous materials present a lower reactivity than the aluminium-rich materials and that the value of the parameter $\triangle$Ro (difference in reactivity in samples dried over P_2O_5 and heated at 105°C) show a different behaviour in aluminium-rich and siliceous allophanic materials, giving positive and negative values, respectively.

These characteristics of the Ro have led us to consider the possibility of using this reaction as a criterion for studying the degree of evolution of the Andisols. In this regard it should be noted that it has been suggested that the andic materials should be defined in therms of active Al rather than amorphous material (WADA 1980, PARFITT 1984).

In the present work, three groups of Andisols in different degrees of evolution are studied, with a view to finding out the behaviour of the parameters deduced from the oxalate reactivity pH 6.3 (Ro; $\triangle$Ro) in each one of them, using the parameters %Al-oxalate pH 3, P retention, Al-pyrophosphate, water retention and bulk density as comparative diagnostic criteria.

ISSN 0722–0723 / ISBN 3–923381–06–9

Tab. 1: CLASSIFICATION OF THE STUDIED SOILS

Profile	SOIL TAXONOMY	ICOMAND 1983	ICOMAND 1984
GROUP 1			
BONA	Hydrandepts	Hapludands	Hapludands
AGUAMANSA	Dystrandepts	Hapludands	Hapludands
LA ESPERANZA	Dystrandepts	Hapludands	Hapludands
TOCHIGI–1 (T–1) (Japan)	Dystrandepts	Hapludands	Hapludands
TOCHIGI–2 (T–2) (Japan)	Dystrandepts	Melanudands	Melanudands
GROUP 2			
CHIO	Vitrandepts	Vitrixerands	Vitrixerands
LAS BREÑAS	Dystrandepts	Vitrudands	Hapludands
GUINATE	⎰Vitrandepts	⎰Vitritorrands	Haploxerands
	⎱Camborthids	⎱Vitrixerands	
ECUADOR–1	Vitrandepts	Vitrudands	Vitrudands
ECUADOR–6	Dystrandepts	Vitrudands	Hapludands
GROUP 3			
LAS RAICES	Dystrandepts	Haplustands	Haplustands
LAS LAJAS	Haplumbrepts	Melanallands	Fulviudands
LAS HAYAS	Haplumbrepts	Melanallands	Melanudands
GARAJONAY	Haplumbrepts	No Andisol	No Andisol

2. MATERIAL AND METHODS

– Soils of the Canary Islands. These are described in the "Guia de Campo" of the International Congress of Volcanic Soils (Tenerife, 1984).

– Soils of Chile and Ecuador. These are described in the book of descriptions of profiles on the VI International Soil Workshop (Chile-Ecuador, 1984).

– Soils of Japan. These are described in the publication corresponding to the "International Correlation of Kuroboku and Related soils" (WADA 1983).

The selected profiles are listed in Table 1, together with their Soil Taxonomy classification and the most recent proposals of Icomand.

Three groups of soils can be identified:

2.1. GROUP 1. ANDIC ANDISOLS

All the soils in this group are found in climatic situations with udic hydric moisture regimes and under different types of vegetation; in the case of the Canary Islands under the Pinar-Fayal Brezal association (La Esperanza) and/or Laurisilva (Aguamansa and Bona). They are typical Andisols; any of the systems of classification, Soil Taxonomy and different proposals of Icomand include them confortably in Andepts and Andisols respectively (Table 1).

2.2. GROUP 2. VITRIC ANDISOLS

Included in this group are the vitric Andisols with a high content of volcanic glass (over

40% in all cases), different degrees of weathering and different climatic situations; the hydric regimes are, respectively: xeric (Chio); udic (Las Breñas, Ecuador–1, Ecuador–6); aridic-xeric (Guinate).

2.3.   GROUP 3. INTERGRADED ANDISOLS

We include in this group those Andisols that present an advanced degree of evolution and weathering towards soils with a much lesser content of amorphous materials.

To differentiate them from the former two groups we call them Intergrades. Within this group all but Garajonay profile can be classified as Andisols because it does not meet the %$\triangle$P requirements.

The soils studied are situated in climates with a udic regime and under Laurisilva and Fayal Brezal vegetation, except in the profile of Las Raíces, with ustic regime, under Pinar. They are found in very geologically old zones, and are associated with ferrallitic-type formations. They possess one or more of the following characteristics: Abundance of kaolinitic clays, Alo values in the andic limits, Alp values of the same order or even greater than those of Alo and P retention that on ocassions does not reach the required limits.

In order to simplify the text we have called the parameters used as follows:

Alo: % of Al extracted with acid ammonium oxalate pH 3 (BLAKEMORE et al. 1981).

Alp: % of Al extracted with 0.1 M sodium pyrophosphate (BLAKEMORE et al. 1981).

$\triangle$P: % Phosphate retention from a potassium dihydrogen phosphate-acetic acid solution, pH 4.6 $\pm$ 0.5, lmg P / ml. (BLAKEMORE et al. 1981).

Ro: Oxalate reactivity pH 6.3. This was carried out on the whole soil in accordance with the technique described in the preceding paper by HERNANDEZ MORENO et al. (1985), always referring to Ro (P_2O_5).

$\triangle$Ro: Difference between the reactivity values determined in a sample dried on P_2O_5 and a sample heated at 105°C (24 h): Ro(P_2O_5) – Ro(105°C).

The allophane content has been calculated according to PARFITT & HENMI (1982) and PARFITT et al. (1983).

3.   RESULTS AND DISCUSSION

In the Andic Andisols (Group 1) the content in allophanic materials is high and consequently the values of Alo, %$\triangle$P and Ro are also high (Table 2).Without exception, the values of Ro are positive, a fact that has been explained in the preceding paper (HERNANDEZ MORENO et al. 1985) and which has been related to the aluminium-rich allophanic materials. It should also be emphasized that those horizons where the differences in water content at 15 bar in moist and dry samples are high give high $\triangle$Ro values.

In the Vitric Andisols (Group 2) the Ro values are relatively low and fluctuate between 5 and 21 mmol/100 g, except for the A_1 and C_1 horizons of the profile Las Breñas. Also, the values of $\triangle$Ro are always negative, a fact associated with allophanic materials with high proportions of silica (HERNANDEZ MORENO et al. 1985).

The A_1 and C_1 horizons of the profile of Las Breñas are closer to the andic Andisols that to the vitric Andisols because of their physical and chemical characteristics: the values of water content at 15 bar, both in dry and in moist samples, are higher that the limit values established in the last Icomand proposal (1984) and the limit value of the Soil Taxonomy in

Tab. 2: ANDIC ANDISOLS

Soils		Alo	Alp	Ro	$\triangle$Ro	%$\triangle$P	BD	pF 15 atm. (dry)	pF 15 atm. (moist)
BONA	A_1	8,3	3,1	96	+ 65	99,6	0,44	37,5	37,7
	Bw_1	12,5	2,6	191	+ 174	99,8	0,42	42,6	78,6
	Bw_2	12,4	2,2	179	+ 163	99,8	0,55	45,3	78,1
	Bw_3	11,1	1,8	183	+ 150	99,8	0,68	48,1	78,1
	Bw_4	6,9	1,8	99	+ 44	99,6	0,57	41,4	64,8
LA ESPERANZA	A_1	4,6	0,6	48	+ 5	98,1	0,64	29,7	33,3
	Bw/C_1	7,9	0,4	81	+ 17	99,2	0,53	42,0	63,9
	Bw/C_2	6,7	0,4	84	+ 19	99,5	0,49	41,5	65,0
AGUAMANSA	A_1	4,9	1,0	39	+ 4	96,6	0,52	23,8	23,4
	Bw_1	5,8	0,4	51	+ 22	97,0	0,51	25,6	25,0
	Bw_2	6,4	0,4	49	+ 13	97,5	0,59	29,2	30,5
	Bw_3	6,3	0,4	60	+ 8	98,0	0,62	30,0	31,9
	2Bw	10,9	0,5	81	+ 33	98,5	<0,85	39,9	58,7
CHILE–007	BA	4,0	1,5	107	+ 62	98,0	0,55	21,3	65,1
	Bs_1	4,1	1,2	91	+ 49	98,0	0,75	19,4	105,2
	Bcs	5,8	0,6	106	+ 63	99,0	0,80	21,2	135,0
CHILE–009	E	6,7	1,4	100	+ 45	99,0	0,68	35,6	67,3
	Bs_1	6,8	1,3	86	+ 44	99,0	1,14	32,3	73,7
JAPAN T–1	IIB_2	46,0*	—	77	+ 30	99,8	<0,85	—	—
JAPAN T–2	A_{11}	44,0*	—	89	+ 38	99,5	<0,85	—	—

Alo: % Al-oxalate at pH 3; Alp: % Al pyrophosphate; Ro: oxalate reactivity pH 6,3;
$\triangle$Ro: Ro(P_2O_5)-Ro(105°C); %$\triangle$P: % P retention (BLAKEMORE 1978); BD: bulk density.
*   soluble fraction in oxalate pH 3,5 (WADA & WADA 1977).

Tab. 3: VITRIC ANDISOLS

Soils		Alo	Alp	Ro	$\triangle$Ro	%$\triangle$P	BD	pF 15 atm. (dry)	pF 15 atm. (moist)
CHIO	A_1	1,7	0,40	15	− 5	61,3	—	8,8	8,7
	C_1	2,1	0,11	13	− 5	70,3	—	7,4	5,7
	C_3	1,0	0,05	5	− 6	49,5	—	6,1	4,8
	C_4	0,6	0,04	5	− 3	36,5	—	4,3	2,5
LAS BREÑAS	A_1	4,7	0,60	34	+ 9	95,4	—	18,8	17,4
	C_1	6,3	0,30	40	+ 7	95,7	—	26,4	17,1
	$2C_2$	4,3	0,20	21	− 19	87,5	—	17,8	16,0
	4C	3,3	0,20	14	− 21	73,8	—	18,8	13,6
GUINATE	Bw/C	1,1	0,10	18	− 9	56,6	—	18,2	—
ECUADOR–1	A_1	0,5	—	6	− 6	49,0	1,25	7,4	8,8
ECUADOR–6	A_1	1,0	—	20	− 8	83,0	1,07	15,5	17,4

For symbols explanation see table 2.

Tab. 4:   ANDISOL INTERGRADES

Soils		Alo	Alp	Ro	$\triangle$Ro	%$\triangle$P	BD	pF 15 atm. (dry)	pF 15 atm. (moist)
LAS LAJAS	A_1	2,2	2,3	46	− 9	95,0	0,63	36,7	38,1
	Bw	1,5	2,2	33	− 12	92,9	0,85	25,3	31,9
LAS HAYAS	A	2,9	2,1	45	− 9	97,7	0,50	25,5	−
GARAJONAY	A	2,3	1,5	25	+ 4	63,0	0,80	23,6	−
LAS RAICES	A_1	2,7	0,5	21	− 3	90,0	0,60	17,5	19,9
	Bw	3,6	0,2	36	0	92,7	0,68	24,7	28,9
	2B	0,5	0,6	9	− 4	61,3	1,17	22,5	22,5

For symbols explanation see table 2.

one horizon (26 and 4%); likewise, the values of P retention and Alo are also typical of andic Andisols. The positive values of $\triangle$Ro agree with these andic characteristics. With regard to the vitric properties, apart from the high content in glass, it should be pointed out that the water content at 15 bar is very close in dry and moist samples. In fig. 1 it can be observed that in these A_1 and C_1 horizons, the reactivity values Ro are relatively low in comparison with the allophane content, which can be explained by the nature of the allophanic material that, although being more aluminium-rich than in the underlying horizons, is still relatively siliceous; for example, the molar ratio Al/Si = 1.48 in the allophane determined by ammonium oxalate pH 3.

It is therefore not surprising that there is no agreement between the classifications of this soil by the different systems used (Table 1). A lack of agreement is also observed in this Table in the case of the Ecuador–6 and Guinate profiles; however, both the diagnostic criteria and the Ro parameters bring them closer to a vitric soil than in the previous case.

The intergraded Andisols (Group 3) are characterised by having a low or nonexistent content of allophane (according to the method used) and a relatively high Ro in comparison with these values (fig. 1). Taking into account the amounts of Al and Fe extracted with pyrophosphate, it can be considered that the active Al and Fe associated with the humus contribute notably to the Ro. In this regard, it should be borne in mind that the complexes of Al and Fe bound to the humus can have a high reactivity, as shown by WADA et al. (1979) using the adsorption of phosphate by these complexes in Andisols.

With regard to the values of $\triangle$Ro, negative values are found to prevail, as in the case of vitric soils. This fact can be attributed, as indicated in the preceding paper (HERNANDEZ MORENO et al. 1985), to an increase in the acidity of the clay minerals (HENMI & WADA 1974, HENMI 1977). In the case of the organo-mineral complexes, a similar effect can also be expected upon heating (MORTLAND 1970), although this problem would require a more in-depth study.

The interest presented by the parameters deduced from the kinetics of the reaction of ammonium oxalate at pH 6.3, Ro and $\triangle$Ro, in the study of Andisols and Intergrades has become apparent in this comparative study. Within the Andisol Order, the lowest values of Ro are obtained in the vitric soils. The andic soils with "hydric" characteristics attain the highest values. The distribution of values in the soils studied is indicated in fig. 2.

In fig. 1, no defined relationship is observed between the Ro values and the % of

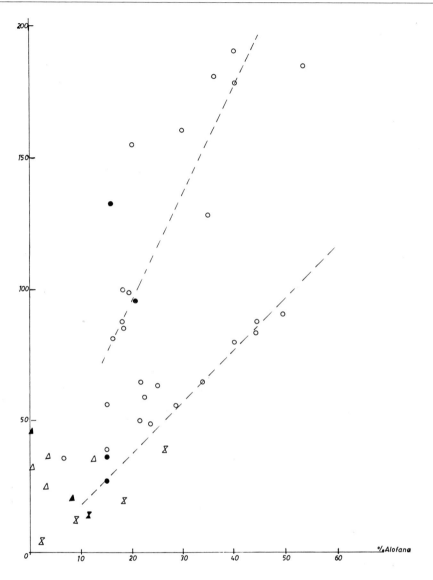

Fig. 1: Ro(mmol/100 g) versus % allophane content (black symbols represent A horizons). The dashed lines only have a reference value. ○ Andic Andisols; ⧖ Vitric Andisols; △ Andisols Intergrades.

allophane. At the same % of allophane it is possible to find very different values of Ro. This may be attributed to the different content of materials of different reactivity. WADA et al. (1979) classify the active aluminium in a decreasing degree of activity: Allophane-like > Al-humus > Allophane, Imogolite. However, PARFITT et al. (1980) considered that the allophane-like constituent could represent the more soluble part of the allophane and imo-

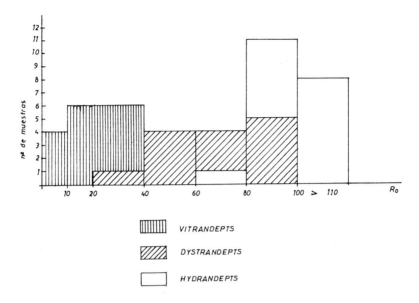

Fig. 2:   Distribution of Ro (mmol/100 g) values in different types of Andisols.

golite, attributable to defects, pores, edges, more than to any one independent constituent.

Fig. 1 allows the following to be differentiated: andic Andisols with more or less reactivity; vitric soils with relatively low Ro values in comparison with their content in allophane;
and intergrades with high Ro values in relation to their content in allophane and where the
organo-mineral complexes can play an important role.

In fig. 3, the values of P retention are represented vs both Alo and Ro (Alo-$\triangle$P and Ro-
$\triangle$P). It is observed that $\triangle$P is more closely correlated with Ro than with Alo. Taking the
values of 95% ($\triangle$P) as the highest limit, from which, approximately, the curve becomes
asymptotic, the resulting polynomic function presents a correlation coefficient of $r = 0.66$ for
the ratio Alo $- \triangle$P and $r = 0.85$ for the Ro $- \triangle$P. This can be attributed to the fact that Ro represents better the surface characteristics of the materials (amount and type of surface).

The values of Ro vs $\triangle$Ro, (the variations undergone by this parameter upon heating at
105°C) are represented in fig. 4. All the andic Andisols are situated to the right. The vitric
samples, with lower reactivity and/or negative values of $\triangle$Ro, are situated to the left, in the
lower zone. The Intergrades can be found between both types.

Fig. 2 and 3 may, therefore, be an excellent working documents for the study of the
properties and evolution of the Andisols. Likewise, they permit a continuum to be established between the vitric and andic Andisols, as members of a single family, irrespective of
their andic and vitric nature, for the study of the evolution of the soil once the Order has been
defined.

Taking into account the simplicity and quickness of the method, we consider its application to a larger number of soils and reference models to be of great interest.

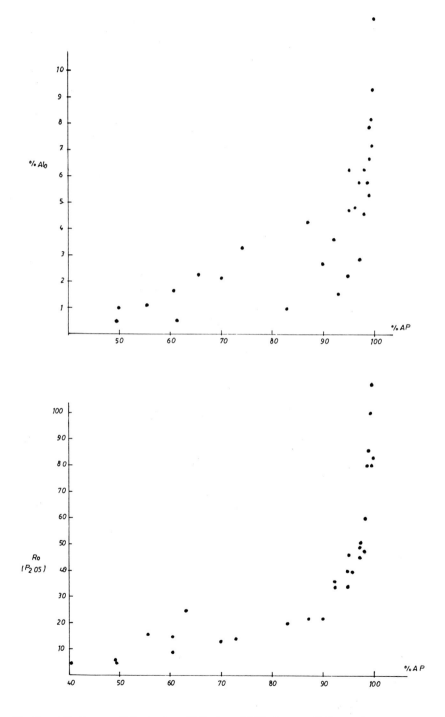

Fig. 3:   a) % Al versus %P retention; b) Ro(mmol/100 g) versus % P retention.

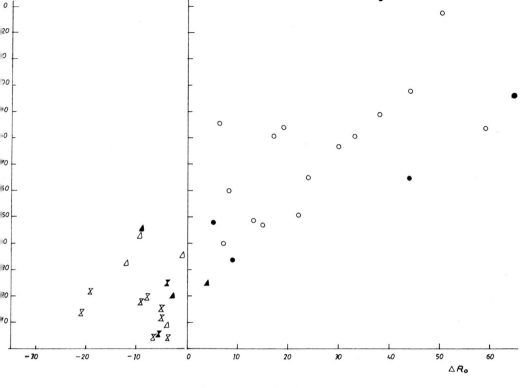

Fig. 4:   Ro(mmol/100 g) versus Ro. (for symbols see fig. 1).

REFERENCES

BLAKEMORE, L.C., SEARLE, P.L. & DALY, B.K. (1981): Methods for chemical analysis of soils. New Zealand Soils Bureau Scientific Report. IDA.

GONZALEZ BATISTA, A., GARCIA, J., HERNANDEZ MORENO, J. y FERNANDEZ CALDAS, E. (1982): Estudio de la cinética de la reacción del oxalato amónico con arcillas alofánicas. Anales de Edaf. y Agrob., T. XLI, Nums. 5-6, 915-926.

HENMI, T. (1977): The dependence of surface acidity on chemical composition of allophane. Clay Min., 12, 356-368.

HENMI, T. & WADA, K. (1974): Surface acidity of Imogolite and Allophane. Clay Min., 10, 231-245.

HERNANDEZ MORENO, J., CUBAS GARCIA, V., BATISTA, A.G. y FERNANDEZ CALDAS, E. (1985): Estudio de la reactividad oxalato pH 6.3 (Ro) en suelos de carga variable. Congreso Internacional de suelos volcánicos. Tenerife, 1984. CATENA S 7: Volcanic Soils. (in press).

ICOMAND. – Circular Letters Nos. 5 (1983) & 6 (1984). N.Z. Soil Bureau.

International Congress of Volcanic Soils. Tenerife, Canary Islands. July, 1984. "Guia de Campo". VI International Soil Classification Workshop. Chile–Ecuador. 1984. Book of descriptions of profiles.

MORTLAND, M.M. (1970): Clay-organic complexes and interactions. Adv. Agron., 22, 75-117.

PARFITT, R.L. (1984): The nature of andic and vitric materials. Congreso Internacional de suelos volcánicos. Tenerife, 1984. Proceedings, 413-435.

PARFITT, R.L. & HENMI, T. (1982): Comparison of an oxalate extraction method and IR method for determining allophane in soils. Soil Sci. Pl. Nut., 28, 183-190.

PARFITT, R.L., RUSSELL, M. & ORBELL, G.E. (1983): Weathering sequence of soils from volcanic ash involving allophane nd halloysite, New Zealand. Geoderma, 29, 41-57.

WADA, K. (1983): International Correlation fo Kuroboku and related soils. Grant–in–Aid for Co-operative Research Nr. 563006 (1981-1982).

WADA, K. (1980): Mineralogical characteristics of Andisols. In: Soils with Variable Charge. B.K.G. Theng (ed.). N.Z., 448.

WADA, K. & GUNJIGAKE, N. (1979): Active Al and Fe, and P sorption in Ando Soils. Soil Sci., **128**, Nr., **6**, 331-336.

Address of authors:
E. Fernández Caldas, J. Hernández Moreno, M.L. Tejedor Salguero, A. González Batista and V. Cubas García
1. Departamento de Edafología, Facultad de Bilogía. Universidad de La Laguna, Carretera Vieja de La Esparanza
La Laguna. Tenerife (Canary Islands). SPAIN

E. Fernandez Caldas & Dan H. Yaalon (Eds):
VOLCANIC SOILS
CATENA SUPPLEMENT 7, Braunschweig 1985

SURFACE CHARGE CHARACTERISTICS OF VOLCANIC ASH SOILS FROM THE SOUTHERN HIGHLANDS OF PAPUA NEW GUINEA

D.J. Radcliffe, Mendi, and G.P. Gillman, Townsville

SUMMARY

The ion exchange properties of six Andept profiles from the Southern Highlands Province of Papua New Guinea were measured at an ionic strength of 0.006 over a range of pH from around 4.0 to 7.0. The point of zero charge of the soil variable charge components was determined, allowing the estimation of actual amounts of permanent and variable charge in the soil. Clear differences were observed between soils developed directly in airfall volcanic ash and those formed in ash which had been alluvially sorted and redeposited. Although total permanent charge was similar in both groups, airfall ash soils showed greater increases in total negative charge with increasing pH, indicating a greater potential variable charge than in alluvial ash soils. In all soils, pH_0 was highly positively correlated with free iron and aluminium oxide content, while for the airfall ash soils, pH_0 was also negatively correlated with organic carbon. The relatively low amounts of negative charge in these soils is discussed in relation to their limited capacity to retain nutrient cations and to future research needs to overcome this limitation.

1. INTRODUCTION

Soils formed in volcanic ash parent materials are the dominant cultivated soils in the central and western highland province of Papua New Guinea. These soils commonly support a semi-permanent agricultural system based on sweet potato *(Ipomoea batatas,* L.) and support some of the highest population densities in the country.

In spite of their extensive present use, highland volcanic ash soils have significant fertility problems which are accentuated by agricultural intensification. An understanding of the factors governing soil fertility is an essential prerequisite to determining the feasibility of ameliorating these constraints and increasing soil productivity to meet the subsistence and cash crop needs of an expanding population.

This paper focusses on the charge characteristics of volcanic ash soils from the Southern Highlands Province of Papua New Guinea. Negative and positive charges associated with soil particle surfaces are essentially responsible for controlling the supply of certain nutrients to the crop plant and for determining the response to soil amendments and fertilizer. The particular clay mineralogy of volcanic ash soils gives rise to specific surface charge characteristics which have only recently been studied in detail. The present account examines the variation in surface charge over a range of pH in selected soil profiles and evaluates the results in terms of causative factors and implications for soil management.

ISSN 0722–0723 / ISBN 3-923381–06–9

2. THEORY

Electric charge at the surface of soil clay particles arises either from isomorphous substitution in the lattice of crystalline clay minerals or from adsorption and/or desorption of protons at the soil-solution interface. The former type of charge is permanent and is unaffected by soil solution conditions. Conversely the charge associated with protonation and deprotonation of the particle surface is highly dependent on the concentration and pH of the contact electrolyte.

The type and amount of soil charge is a function of the clay mineralogy and organic matter content. Volcanic ash soils typically have clay mineralogies dominated by short range order minerals such as allophane and ferrihydrite, have high organic matter contents, and are extreme examples of variable charge soils. Classical double layer theory can be applied to describe the charge distribution in a variable charge soil system as demonstrated by VAN RAIJ & PEECH (1972). UEHARA & GILLMAN (1981) used Gouy-Chapman and Nernst theory to arrive at the following equation describing the relationship between charge and solution conditions.

$$\sigma_V = \left(\frac{2 \, n \, \varepsilon \, k \, T}{\pi} \right)^{1/2} \sinh 1.15 \, z \, (pH_0 - pH) \qquad (1)$$

where σ_V = net variable surface charge, n is the concentration of electrolyte solution of dielectric constant ε, k is the Boltzmann constant, T is the absolute temperature, z is the counterion valance and pH_0 is the pH at which the net variable charge is zero.

Equation (1) shows that in a variable charge system the sign of the net charge is determined by the solution pH relative to pH_0. Furthermore once the sign of the surface charge is fixed by pH, its magnitude is then governed by the difference between pH and pH_0 and by the composition and concentration of the contacting electrolyte solution.

Even in soils dominated by short range order clay minerals a small amount of permanent charge may be present and the total charge in the soil system can be represented by

$$\sigma_T = \sigma_V + \sigma_P \qquad (2)$$

where σ_T is the total charge and σ_P is the net permanent charge.

Figure 1 illustrates typical negative and positive charge curves at constant electrolyte concentration in a mixed permanent-variable charge system, and shows the importance of pH_0 in defining the charge components. The net charge at pH_0 will be permanent charge and the sign and magnitude of σ_P can be estimated as shown in the example. In a system with net permanent negative charge, pH_0 will be higher than the point of zero net charge (PZNC) where the total negative and total positive charges are equal ($\sigma^-_T = \sigma^+_T$).

3. EXPERIMENTAL

3.1. DESCRIPTION OF SITES AND SOILS

Profiles representative of six soil series were selected for analysis. Table 1 gives details

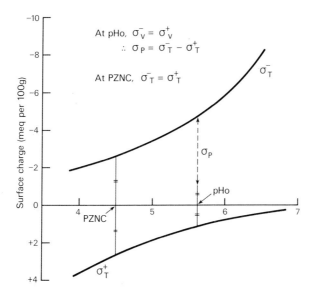

Fig. 1:   Typical curves for negative and positive charge showing the positions of pH_0, PZNC and how σ_P is estimated.

Tab. 1:   SITE CHARACTERISTICS AND SOIL CLASSIFICATION

Site No.	Location	Grid Ref*	Elevation (m)	Annual** Rainfall (mm)	Parent Material	Soil Classification		
						USDA	ICOMAND	Series
49	Pomiorine Exp. Farm	AN 817941	1550	4864	Airfall Tephra	Typic Hydrandept	Hapludand	Pangia
9	Kiburu Exp. Farm	YU 948173	1700	2887	Airfall Tephra	Typic Hydrandept	Hydrudand	Kalongia
329	150 m W Tura Vill.	YU 899269	1850	2887	Airfall Tephra	Typic Hydrandept	Hydrudand	Was
333	550 m SSE Karil Vill.	ZU 017311	2270	3000-4000	Airfall Tephra	Typic Hydrandept	Hydrudand	Birop
31	Piwa Exp. Farm	YU 181517	1620	2542	Alluvium	(Aquic) Entic Eutrandept	Hapludand	Piwa
3	Kiburu Exp. Farm	YU 947172	1680	2887	Alluvium	Oxic Dystrandept	–	Kiburu

*   From Papua New Guinea 1:100,000 Topographic Survey Sheets.
** Median annual rainfall, at Mendi (1952-1983) (6°09'S; 143°39'E; 1875 m) for sites 9, 329, 3; at Pangia (1962-1977) (6°21'S; 144°07'E; 1620 m) for site 49 and at Tari (5°52'S; 142°SS'E; 1610 m) for site 31. Estimated value for site 333.

on the location and environment of the sites and the classification of the soils. Three of the profiles, representing the Kalongia, Was, and Birop Series are formed in airfall volcanic ash from a common source, probably corresponding to the Tomba tephra described by PAIN & BLONG (1976). The Pangia Series is formed in airfall tephra from a different source and the Piwa and Kiburu Series are formed in ash which has been alluvially sorted, mixed with residual materials and redeposited. The soil moisture regime is perudic at all sites and the soil temperature regime is isothermic for all the sites except 333 where it is isomesic.

The soils are classified according to Soil Taxonomy (SOIL SURVEY STAFF 1975) and the most recent ICOMAND* circular, in Table 1. In the original Soil Taxonomy all the profiles are classified as Andepts, and sites 49, 9, 329 and 333 show the irreversible dehydration properties of Hydrandepts. The remaining alluvial profiles are either Dystrandepts or Eutrandepts depending on whether the base saturation exceeds 50%, a distinction which is rather meaningless in soils of dominantly variable charge. According to the current proposals of ICOMAND one of the alluvial profiles (3) fails to qualify as Andisol on the criteria for oxalate extractable aluminium. Three of the airfall ash soils (9, 329 and 333) qualify as Hydrudands and the remaining two profiles (31 and 49) are Hapludands.

3.2. ANALYTICAL METHODS

The procedure described by GILLMAN (1984) was used for the measurement of negative and positive charge between pH4 and pH7 when the soil was in equilibrium with 0.002 M $CaCl_2$ solution. Procedures for determining pH_0 and soil pH were also described by GILLMAN (1984). In the determination of surface charge and pH_0, any sorbed sulphate was removed by washing the soil with 0.1 M $CaCl_2$ with pH adjusted to about 10. Free Fe_2O_3 and Al_2O_3 were extracted with a mixture of 1% sodium dithionite and 5% sodium citrate (HOLM-GREN 1967) and organic carbon was determined by the method of WALKLEY & BLACK (1934). Exchangeable basic cations were extracted with a mixture of 0.1 M $BaCl_2$ and 0.1 M NH_4Cl, while acidic cations were extracted with 1 M KCl and the extract titrated to pH 8.5. The compulsive exchange method of GILLMAN (1979) was used to measure cation exchange capacity.

4. RESULTS AND DISCUSSION

Available data on the 'free' iron and aluminium oxide, organic carbon, and exchangeable cation content of the samples anaylsed are summarized in Table 2. Limited available data (R.F. ALLBROOKE, Waikato University, pers. comm.) indicate that the clay fraction is allophane and ferrihydrite dominant with, particularly in the alluvial soils, smaller amounts of halloysite and gibbsite. Profile descriptions and full analytical data are available in the Technical Reports of the Southern Highlands Rural Development Project (RAD-CLIFFE 1983a, 1984a, 1984b, 1984c).

The variation in negative and positive charge over a range of pH is illustrated for the soil series and component horizons in Figure 2. The results are analysed with the aims of:
–   defining the charge components at pH_0.
–   comparing the steepness of the charge curve and the change in charge between standard pH values.
–   defining the charge components and resulting cation and anion exchange capacities at soil pH.

Table 3 lists the total negative and positive charge (σ^- and σ^+) at pH_0. As the net variable charge at pH_0 is zero, any residual charge is net permanent charge (σ_p). pH_0 invariably

---

*   International Committee on the Classification and Management of Andisols Circular No. 6.

Tab. 2: FREE SESQUIOXIDES, ORGANIC CARBON, AND EXCHANGEABLE CATION CONTENT.

Soil Series	Horizon and Depth (cm)	Fe_2O_3 (%)	Al_2O_3 (%)	Organic Carbon (%)	Ca	Mg	K	Na	H+Al
						(meq per 100 g)			
Pangia	A	0-15	11.7	6.6	7.8	1.6	0.5	0.14	0.10	1.3
	AB	15-19	12.2	7.3	5.5	0.9	0.3	0.13	0.11	0.3
	B	19-50	16.5	8.6	2.6	0.6	0.2	0.04	0.10	0.1
Kalongia	A	0-21	10.5	7.6	9.8	1.7	0.2	0.14	0.17	0.1
	B1	21-40	13.9	8.5	5.2	0.7	0.1	0.07	0.12	0.1
	B2	40-60	21.5	11.3	4.2	0.5	0.1	0.03	0.08	0.1
Was	A1	0-15	6.0	6.1	12.8	13.8	2.0	0.23	0.17	0.6
	A2	15-32	6.2	6.4	13.1	12.3	2.0	0.20	0.18	0.8
	B	32-55	19.2	11.9	3.9	0.6	0.1	0.06	0.12	0.1
Birop	A1	0-15	7.5	6.7	14.8	3.0	0.6	0.29	0.13	1.8
	A2	15-30	9.9	8.0	10.9	1.6	0.2	0.16	0.13	1.1
	C	85-125	0.3	1.5	1.0	0.1	0.1	0.05	0.11	0.1
Piwa	A	0-17	2.0	2.6	3.2	2.6	0.5	0.07	0.15	0.7
	B1	17-50	1.0	1.4	1.4	2.1	0.2	0.09	0.26	0.5
	B2	50-80	0.5	1.1	0.5	8.4	1.5	0.16	0.42	2.2
Kiburu	A1	0-15	9.8	4.8	4.3	6.6	0.6	0.08	0.11	0.5
	A2	15-30	10.1	4.8	3.9	5.8	0.6	0.05	0.14	0.6
	B	30-60	11.2	4.3	0.8	5.4	1.3	0.05	0.13	0.4

shows a marked increase from the A horizons to the subsoil horizons in the airfall ash soils while the alluvial soils have consistently low values. With the marginal exception of the Piwa B1 horizon and the Birop C horizon all the soils have net negative permanent charge thoughout the profile although there is a consistent decrease in the magnitude of σ_p from the A horizons to the subsoil horizons in the airfall ash soils. This trend is a product of both decreasing σ^- and increasing σ^+ from the topsoil to the subsoil

Figure 2 illustrates the variation in total negative and positive charge with pH. Regression equations of the form

$$\sigma_T = ae^{b(pH_0-pH)} \tag{3}$$

fitted to the curves accounted for between 88% and 100% of the variation in negative charge and between 71% and 99% of the variation in positive charge with pH. The factor 'a' in Eq. 3 corresponds to the charge at pH_0 (Table 3), while 'b', the rate coefficient, describes the steepness of the exponential curve.

Table 4 summarizes the variation of charge with pH by comparing the rate coefficients and by showing the charge at standard pH values of 4.0 and 7.0. The variability of negative charge is significantly greater in the airfall ash profiles as opposed to the alluvial profiles and the former soils show a consistent increase in rate coefficient from the A horizons to the subsoil horizons. Positive charge curves have similar rate coefficients in both soil groups and show a consistent increase in steepness with depth in the profile. The contrasting rate coefficients of the charge curves reflect the comparative differences in σ^- and σ^+ measured at the

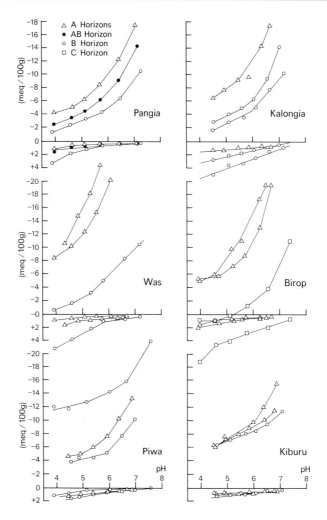

Fig. 2: Negative and positive charge curves for the six soils studies.

standardized pH values. In the A horizons of the airfall ash soils σ^- at pH7 is 4 to 5 times σ^- at pH4 while the increase is 5 to 8 times in the B horizons. The single C horizon, from the high altitude Birop Series, shows the most extreme variable charge properties and the positive values of the negative charge curve at low pH (Figure 2) are due to 'negative adsoprtion' or repulsion of cations from the highly positive charged surface. Alluvial soils show lesser increases in negative charge (1.5 to 4 times) between pH4 and pH7. All the samples tested showed net negative charge at pH7 and the alluvial soils and airfall ash A horizons also have a net negative charge at pH4. Conversely all the airfall ash subsoils display net positive charge at pH4 indicating that this pH is not only below pH_0 (Table 3) but is also below PZNC (Figure 1).

Charge characteristics at soil $pH\left(pH_{CaCl_2}\right)$ are summarized in Table 5. Values for

Tab. 3: SURFACE CHARGE CHARACTERISTICS AT pH_0

Soil and Horizon		pH_0	σ^-	σ^+ (meq/100 g soil)	σ_p
Pangia	A	4.7	5.5	0.7	−4.8
	AB	4.6	3.4	1.0	−2.4
	B	5.5	3.8	1.0	−2.8
Kalongia	A	4.5	6.2	1.4	−4.8
	B1	5.2	4.2	2.0	−2.2
	B2	6.0	4.8	2.3	−2.5
Was	A1	4.0	9.1	1.8	−7.3
	A2	4.1	8.4	0.8	−7.6
	B	5.7	4.4	1.4	−3.0
Birop	A1	4.3	5.7	1.1	−4.6
	A2	4.6	6.0	1.3	−4.7
	C	6.2	1.2	1.9	+0.7
Piwa	A	3.7	3.1	2.0	−1.1
	B1	3.8	2.7	2.8	+0.1
	B2	3.1	9.7	4.0	−5.7
Kiburu	A1	4.3	6.0	0.8	−5.2
	A2	4.3	5.6	0.9	−4.7
	B	4.1	5.8	2.0	−3.8

Tab. 4: VARIATION OF SURFACE CHARGE WITH pH

Soil and Horizon		Negative Charge			Positive Charge		
		Rate	pH4	pH7	Rate	pH4	pH7
		Coefficient	(meq/100 g soil)		Coefficient	(meq/100 g soil)	
Pangia	A	0.47	4.0	16.2	0.32	0.9	0.3
	AB	0.56	2.4	13.0	0.51	1.4	0.3
	B	0.59	1.6	9.2	0.73	2.9	0.3
Kalongia	A	0.44	5.0	18.8	0.31	1.7	0.6
	B1	0.62	2.0	12.7	0.62	4.2	0.7
	B2	0.68	1.2	9.4	0.59	7.4	1.2
Was	A1	0.51	9.1	42.7	0.55	1.8	0.4
	A2	0.44	8.0	29.9	0.49	0.8	0.2
	B	0.63	1.5	9.9	0.78	5.4	0.5
Birop	A1	0.56	4.8	25.8	0.40	1.3	0.4
	A2	0.47	4.6	18.7	0.68	2.0	0.3
	C	1.32	− 0.9	5.4	0.63	7.6	1.2
Piwa	A	0.43	3.5	12.8	0.56	1.6	0.3
	B1	0.39	2.9	9.3	0.64	2.5	0.4
	B2	0.16	11.2	18.3	1.11	1.5	0.0
Kiburu	A1	0.24	5.9	12.0	0.27	0.9	0.4
	A2	0.37	5.0	15.3	0.21	1.0	0.5
	B	0.22	5.7	11.0	0.67	2.2	0.3

Tab. 5:   SURFACE CHARGE PROPERTIES AT SOIL pH (pH $_{CaCl_2}$).

Soil and Horizon		pH $_{CaCl_2}$	Neg.	Charges at pH $_{CaCl_2}$ Pos. (meq per 100 g)	Variable Neg. Charge at pH $_{CaCl_2}$ (%)
Pangia	A	4.7	5.5	0.7	13
	AB	4.9	4.0	0.8	40
	B	5.0	2.8	1.4	0
Kaongia	A	4.6	6.5	1.4	26
	B1	5.0	3.7	2.3	41
	B2	5.4	3.2	3.2	22
Was	A1	5.1	16.0	1.0	54
	A2	5.0	12.4	0.5	39
	B	5.3	3.4	2.0	12
Birop	A1	4.7	7.1	1.0	35
	A2	4.8	6.6	1.1	29
	C	5.7	1.1	2.6	36
Piwa	A	4.7	4.8	1.1	77
	B1	5.0	4.2	1.3	100
	B2	4.9	12.9	0.5	56
Kiburu	A1	4.8	7.1	0.7	27
	A2	4.8	6.8	0.8	31
	B	4.7	6.6	1.4	42

Tab. 6:   ESTIMATION OF CEC BY THREE METHODS

Soil and Horizon		σ^- at pH $_{CaCl_2}$	CEC by Comp. Exch. meq per 100 g	ECEC
Pangia	A	5.5	5.5	3.6
	AB	4.0	4.9	1.7
	B	2.8	4.4	1.0
Kaongia	A	6.5	6.3	2.2
	B1	3.7	5.4	0.9
	B2	3.2	3.9	0.6
Was	A1	16.0	14.9	16.8
	A2	12.4	14.7	15.5
	B	3.4	4.5	1.0
Birop	A1	7.1	7.5	5.8
	A2	6.6	6.0	3.2
	C	1.1	2.8	0.3
Piwa	A	4.8	4.5	4.0
	B1	4.2	3.9	3.1
	B2	12.9	11.6	12.7
Kiburu	A1	7.1	7.8	7.9
	A2	6.8	10.1	7.2
	B	6.6	4.9	7.3

negative charge are much lower than CEC values obtained for volcanic ash soils especially when using solutions of high pH and electrolyte concentration. The relative importance of positive charge at soil pH should also be noted. In the final column, the proportion of negative charge attributable to the variable charge components at soil pH has been calculated by subtracting permanent charge from total negative charge at soil pH and expressing the result as a proportion of the latter. The effect of altering the pH of the soil on the development of negative and positive charge, is evidenced by comparing the data in Table 5 with that in Table 4. In many cases, variable charge constitutes over 30% of the total negative charge, but this value would be greatly reduced by lowering soil pH.

Estimates of CEC under conditions resembling those in the field have been attempted by three separate methods viz. (1) measuring total negative charge in an unbuffered divalent cation solution at low ionic strength, as already discussed, (2) the compulsive exchange method of GILLMAN (1979), and (3) the Effective Cation Exchange Capacity (ECEC). obtained by summation of basic cations extracted with 0.1 M $BaCl_2$/NH_4Cl and acidic cations extracted with 1 M KCl. Results are tabulated in Table 6. The agreement between the three methods is generally good, but there are some notable exceptions especially in the lower horizons of the airfall ash soils. The reason for the discrepancies is not clear but it is possible that extra specific surface is created in these highly dispersive soils in the first two methods which involve lengthy sample shaking times to achieve equilibrium. This aspect of determining charge properties of andic materials requires further study.

5. GENERAL DISCUSSION

The results summarized in Tables 3 to 5 indicate how negative and positive charge varies with pH under conditions of constant electrolyte composition and concentration. All the soils analysed have substantial variable charge components, with the airfall ash soils showing greater charge response to changing pH than the soils formed in alluvially redeposited ash. The airfall ash soils display clear trends in the factors affecting charge distribution, showing marked increases with depth in pH_0, rate coefficients of the charge curves, and anion exchange capacity (AEC), and corresponding decreases in permanent negative charge, and CEC. No such trends are apparent in the alluvial soils.

The present discussion focusses on the causes of charge distribution in these volcanic ash soils, and considers the implications of these charge properties for soil management.

As stated previously, soil surface charge is determined by clay mineralogy and organic matter content. In cases where short range order minerals dominate the clay fraction the content of 'free' or 'active' sesquioxides and organic carbon may be expected to exert the dominant influence on surface charge. Organic matter has a pH_0 of around 4 whereas values for iron and aluminium oxides are in the range 7 to 9 (UEHARA & GILLMAN 1981, 46) and the tendency of these substances to have antagonistic effects on pH_0 has been described by VAN RAIJ & PEECH (1972) and UEHARA & GILLMAN (1981). Table 7 presents the results of correlations of Fe_2O_3, Al_2O_3 (extracted in sodium citrate/dithionite solution), $R_2O_3(Fe_2O_3 + Al_2O_3)$, and organic carbon with pH_0 and negative and positive charge at both pH_0 and pH $CaCl_2$ for the airfall ash soils.

pH_0 is the attribute most strongly correlated with both sesquioxides and organic carbon. The linear regression lines of pH_0 on organic carbon and free Fe_2O_3 are illustrated in Figure 3. The coefficient of determination (r^2) of 0.97 for Fe_2O_3 could not be increased by including organic carbon in a multiple linear regression. In these airfall ash soils, organic car-

Tab. 7:   CORRELATION COEFFICIENTS OF pH_0 AND CHARGE PARAMETERS WITH ORGANIC CARBON AND FREE SESQUIOXIDE CONTENT IN AIRFALL ASH SOILS

Charge Parameter	Org C (%)	Fe_2O_3 (%)	Al_2O_3 (%)	R_2O_3 (%)
pH_0	− 0.90***	0.98***	0.92***	0.98***
$\sigma^-\ pH_0$	0.84***	− 0.71*		
$\sigma^+\ pH_0$	0.12 n.s.	0.23 n.s.	0.28 n.s.	
σ_p	0.86***	− 0.79**	− 0.43*	
CEC	0.82**	− 0.78**		
AEC	− 0.66*	0.82**	0.82**	0.83**

Note: Birop C horizon not included in sesquioxide correlations.
*** Significant at 0.1% level; ** Significant at 1% level; * Significant at 5% level; n.s. Not significant.

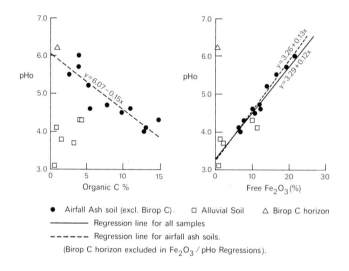

● Airfall Ash soil (excl. Birop C).   □ Alluvial Soil   △ Birop C horizon
────── Regression line for all samples
‐‐‐‐‐‐ Regression line for airfall ash soils.
(Birop C horizon excluded in Fe_2O_3 / pHo Regressions).

Fig. 3:   Relationship between (a) pH_0 and oganic carbon and (b) pH_0 and free Fe_2O_3.

bon and free Fe_2O_3 are highly negatively correlated.

For comparison the data for the alluvial soils are also included in Figure 3. Free Fe_2O_3 accounts for most of the variability in pH_0 with the exception of the Birop C horizon where sesquioxide content is very low. pH_0 is unaffected by organic carbon content in the alluvial soils, which have a more varied suite of clay minerals. The evidence of GONZALES–BATISTA et al. (1982) suggests that the very low pH_0 in Piwa B2 may be related to the relative importance of halloysite in this horizon.

Organic carbon shows the most highly significant correlations with negative charge, both at pH_0 and at soil pH (CEC), which is consistent with the fact that organic matter and clay-organic complexes provide most of the cation exchange sites in soils that are lacking or deficient in 2:1 clay minerals. AEC shows a weak negative correlation with organic carbon but is more strongly correlated with both iron and aluminium oxides.

The presence of substantial amounts of permanent negative charge (Table 3), usually accounting for more than half the total negative charge at soil pH, suggests some isomorphous substitution and, by inference, some 2:1 clay minerals. While the possibility of some crystalline clay minerals, even in the airfall ash soils, cannot be discounted (WOOD (1984) and BLEEKER & HEALY (1980) have detected 14 Å minerals in similar soils from highland Papua New Guinea), the data of PARFITT & MAVO (1975) support the view that 2:1 minerals are virtually absent from soils of this type. According to ESPINOZA et al. (1975) isomorphous substitution can occur in the lattice vacancies of short range order minerals and this process could explain the permanent charge in the soils of the present study.

The implication of the charge properties of the Southern Highlands volcanic ash soils to their management and productive potential depend on the extent to which these properties are limiting present or future crop performance. If the charge characteristics are a constraint to crop production, to what extent can this constraint be ameliorated and what practices can be adopted to minimise its effect on crops?

The results in Table 5 show that, with the exception of the Was Series topsoil and the poorly drained Piwa subsoil, all the samples have limited ability to retain cations at soil pH and this explains the low cation contents recorded in Table 2. It is reasonable to suppose that the limited surface negative charge is limiting crop nutrient supply and the efficiency of fertilizer response. The feasibility of raising CEC therefore requires investigation, and consideration of the charge curves (Figure 2) would suggest that this might be achieved by increasing the negativity of $(pH_0 - pH)$, either by raising the soil pH or decreasing pH_0.

The increase in negative charge resulting from a 0.5 unit increment in pH was calculated from regression equations fitted to the negative charge curves for the four airfall ash soils. A moderate increase of between 1.5 and 4.6 meq/100 g is predicted, raising CEC to 7-10 meq/100 g in the Pangia, Kalongia and Birop Series. Should $(pH_0 - pH)$ become more positive by an equivalent 0.5 unit CEC values would fall to 4-6 meq/100 g.

Raising the pH of highly variable charge soil by liming is rarely feasible because of the high buffering capacity as pH departs from pH_0 (UEHARA & GILLMAN 1981, 47). Furthermore the addition of large amounts of calcium to a system in which potassium is only marginal would exacerbate K deficiency, particularly in tuber crops like the sweet potato staple.

Lowering pH_0 depends on blocking some of the positively charged sites by adsorbed anions. The dependence of pH_0 on organic carbon content (Table 7) in airfall ash soils would suggest that additions of organic matter may be effective in lowering pH_0 (UEHARA & GILLMAN 1981) although the scope for incorporating further organic matter in soils which already have an average topsoil organic carbon content of 11.3% (airfall soils) may be limited. Phosphate and silicate are probably the most promising inorganic anions for lowering pH_0 and thus increasing CEC. As the Southern Highland soils, in common with many other volcanic ash soils, are very deficient in phosphorous due to acute fixation (RADCLIFFE 1983b) phosphate added as a plant nutrient may have additional benefits in raising the CEC.

The technical and economic feasibility of achieving these increases in CEC requires further investigation. Even if increasing the present CEC is not feasible, current levels of exchangeable cations should be closely monitored to ensure there is no further decline as a result of decreasing organic matter levels under increasing cropping intensity or diversification into more demanding crops (WOOD 1984). Given the existing situation of limited CEC, effectiveness of split versus single applications of cationic fertilizers should be checked in the field and the considerable AEC at soil pH in the airfall ash subsoils could be exploited by using fertilizers in forms where the anions are specifically adsorbed (e.g. AYRES & HAGIHARA 1953). In an agricultural system which is primarily subsistence based, the effect of organic materials on CEC as well as on direct nutrient supply requires thorough investi-

4

gation. The results in this paper are seen as an essential step in understanding the processes governing cation and anion retention and as a basis for designing improvements and assessing the productivity of a group of soils that have considerable economic importance in Papua New Guinea.

BIBLIOGRAPHY

AYRES, A.S. & HAGIHARA, H.H. (1953): Effects of the anion on the sorption of potassium by some humic and hydrol humic latosols. Soil Science **75**, 1-17.
BLEEKER, P. & HEALY, P.A. (1980): Analytical data of Papua New Guinea Soils. CSIRO Australia Division of Land Use Research Technical Paper No. **40**, 2 volumes, CSIRO, Melbourne.
ESPINOZA, W., GAST, R.G. & ADAMS, R.S. (1975): Charge characteristics and nitrate retention by two andepts from South-Central Chile. Soil Science Society of America Proceedings **39**, 842-846.
GILLMAN, G.P. (1979): A proposed method for the measurement of exchange properties of highly weathered soils. Australian Journal of Soil Research **17**, 129-139.
GILLMAN, G.P. (1984): Using variable charge characteristics to understand the exchangeable cation status of oxic soils. Australian Journal of Soil Research **22**, 71-80.
GONZALES–BATISTA, A., HERNANDEZ–MORENO, J.M., FERNANDEZ–CALDAS, E. & HERBILLON, A.J. (1982): Influence of silica content on the surface charge characteristics of allophanic clays. Clays and Clay Minerals **30**, 103-110.
HOLMGREN, G.G.S. (1967): A rapid citrate-dithionite extractable iron procedure. Proceedings of the Soil Science Society of America **31**, 210-211.
PAIN, C.F. & BLONG, R.J. (1976): Late quaternary tephras around Mt Hagen and Mt Giluwe, Papua New Guinea. In: Volcanism in Australia. (Ed. R.W. Johnson). Elsevier, Amsterdam.
PARFITT, R.L. & MAVO, B. (1975): Phosphate fixation in some Papua New Guinea soils. Science in New Guinea **3**, 179-190.
RADCLIFFE, D.J. (1983a): Land resources of Pomiorine Experimental Farm. AFTSEMU Technical Report No. 3. Southern Highlands Rural Development Project. Mendi, Papua New Guinea.
RADCLIFFE, D.J. (1983b): The management properties of Andisols in Southern Highlands Province, Papua New Guinea. AFTSEMU Technical Report No. 1. Southern Highland Rural Development Project, Mendi, Papua New Guinea.
RADCLIFFE, D.J. (1984a): The land resources of Kiburu Experimental Farm. AFTSEMU Technical Report No. 5. Southern Highlands Rural Development Project, Mendi, Papua New Guinea.
RADCLIFFE, D.J. (1984b): The land resources of Piwa Experimental Farm. AFTSEMU Technical Report No. 6. Southern Highlands Rural Development Project, Mendi, Papua New Guinea.
RADCLIFFE, D.J. (1984c): The land resources of the upper Mendi area. AFTSEMU Technical Report No. 8. Southern Highlands Rural Development Project, Mendi, Papua New Guinea.
SOIL SURVEY STAFF (1975): Soil taxonomy. A basic system of soil classification for making and interpreting soil surveys. United States Department of Agriculture Handbook No. 436. (Govt. Printing Office: Washington, D.C.).
UEHARA, G. & GILLMAN, G.P. (1981): The mineralogy, chemistry, and physics of tropical soils with variable charge clays. Westview Press, Boulder, colorado, 1981. 170 pp.
VAN RAIJ, B. & PEECH, M. (1972): Electrochemical properties of some oxisols and alfisols of the tropics. Soil Science Society of America Proceedings **36**, 587-593.
WALKLEY, A. & BLACK, I.A. (1934): An examination of the Degtjareff method for determining soil organic matter, and a proposed modification of the chromatic acid titration method. Soil Science **37**, 29-38.
WOOD, A. (1984): Land for tomorrow. Subsistence agriculture, soil fertility and ecosystem stability in the New guinea Highlands. Ph.D. Thesis, University Papua New Guinea, Port Moresby.

Addresses of authors:
D.J. Radcliffe, Southern Highlands Rural Development Project, P.O. Box 98
Mendi, S.H.P. Papua New Guinea.
Present Address: Halcrow Fox (SSTD), Kotak Pos 220, Palembang, Indonesia.
G.P. Gillman, CSIRO Division of Soils, Private Mail Bag, P.O.
Aitkenvale, Townsville, Qld. 4814, Australia

E. Fernandez Caldas & Dan H. Yaalon (Eds):
VOLCANIC SOILS
CATENA SUPPLEMENT 7, Braunschweig 1985

KINETIC STUDY OF THE EXPERIMENTAL WEATHERING OF AUGITE AT DIFFERENT TEMPERATURES

J. **González Bonmatí**, M.P. **Vera Gómez**, J.E. **García Hernández**, La Laguna

SUMMARY

An experimental study of the weathering of augite of optimum crystallinity and purity has been carried out at 25, 32 and 40°C by applying the kinetic model of weathering in a flow system. Kinetic curves of dissolution at these temperatures for each of the elements composing the crystal network are obtained, together with related parameters: mechanism of dissolution (surface-dissolution), discussion of congruence, evolution of congruence, evolution of the interface of transfer and others.

The study of the relationship between the initial rates of dissolution and temperature permits the activation energy ($\triangle G^{\mp}$) to be determined, by applying the Arrhenius equation.

The study of the limiting value of the dissolution at 25°C and 1 atm, when $t \rightarrow \infty$, has enables the equilibrium state to be characterised by the determination of the solubility product ($K^0{}_{sp}$) and the free energy of formation ($\triangle G^0{}_f$) of augite, under standard conditions.

RESUMEN: ESTUDIO CINETICO DE LA ALTERACION EXPERIMENTAL DE LA AUGITA A DIFERENTES TEMPERATURAS

Se ha estudiado experimentalmente la alteración de una augita, de cristalinidad y pureza óptimas a 25, 32 y 40°C, por aplicación del modelo cinético de alteración en sistema de percolación, lo que nos ha permitido obtener las curvas cinéticas de disolución a estas temperaturas para cada uno de los elementos constitutivos de la red cristalina y parámetros relacionados: mecanismos de disolución (superficie-difusión), discusión acerca de la congruencia, evolución de la interfase de transferencia, etc.

El estudio de la relación entre las velocidades iniciales de disolución con la temperatura, permite extraer la energía de activación ($\triangle G^{\mp}$), por aplicación de la ecuación de Arrhenius.

El estudio del valor límite de la disolución a 25°C y 1 atm., cuando $t \rightarrow \infty$, nos ha permitido cuantificar el estado de equilibrio y extraer los valores de la constante del producto de solubilidad ($K^0{}_{sp}$) y la energía libre de la formación de la augita ($\triangle G^0{}_f$), en condiciones standard.

1. INTRODUCTION

The rate of dissolution of silicates as a function of temperature (T), has been studied by many authors: DELMAS & PEDRO (1971), HELGESON (1971), BAILEY (1971, 1974), LUCE et al. (1972) and GRANDSTAFF (1977) among others. These studies permit an analysis to be made of the evolution of the mecanism of dissolution as a function of T, together with an estimation of the activation energy ($\triangle G^{\mp}$) for the dissolution reaction. This study looks specifically at augite.

The dissolution of augite and similar pyroxenes has been studied by KELLER et al. (1963), HUANG & KELLER (1970, 1972) and GRANDSTAFF (1977), among others. They

ISSN 0722–0723 / ISBN 3–923381–06–9

demonstrated that solution of augite is incongruent, which is hardly surprising, in view of the chemical and structural complexity of the mineral.

The kinetic model of experimental weathering, applied so far to the study of olivine and cristobalite, provides data for a general scheme for the kinetic interpretation of the reactions of the type crystalline silicates-solution. Such reactions proceed by mixed mechanisms of dissolution, each mechanism controlling the reaction for a given period of time, and showing a change from surface control to diffusion control as time proceeds: GARCIA HERNANDEZ (1981, 1982) and DELMAS et al. (1982).

Here we study the system augite-H_2O through transitory metastable states to a postulated state of pseudo-equilibrium as time approaches ∞. The method has been used before in the study of cristoballite-H_2O ERNST & CALVERT (1963), GARCIA HERNANDEZ (1981), and confirmatory field evidence has been provided TEJEDOR SALGUERO et al. (1978).

Our work allows us to obtain a solubility product (K_{sp}) at 25°C and 1 atm. From this the standard free energy of dissolution ($\triangle G^0_{diss}$) is determined and hence a standard free energy of formation for augite ($\triangle G^0_f$).

A $\triangle G^0_f$ for augite has been determined before (HUANG & KELLER 1972) but the pH was 9.2, where the mechanism of solution was different, involving charged Si species as well as $Si(OH)_4$.

2. MATERIALS AND METHODS

The experimental apparatus consists of a system of 10 teflon reactors, 1.9 cm. interior

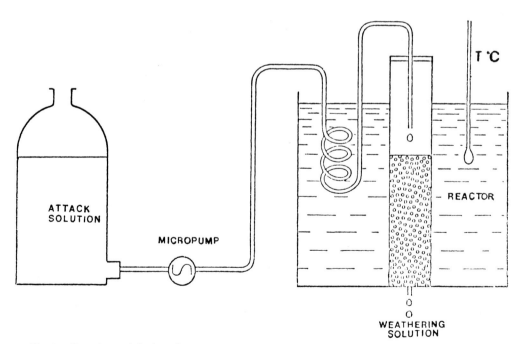

Fig. 1: Experimental device scheme.

diameter and 25 cm. high, containing the mineral under study. The system is thermostated at a working temperature of $\pm 0.1°C$ (Figure 1).

The attack solution consists of pure H_2O, twice demineralized in a Maxy-Recherche multiple layer column, and with a conductivity of $\sim 10^{-6}$ siemens and a pH $= 6.03$. The attack solution is pumped by a series of mechanical micropumps from MPL Metering Pumps Ltd., regulatable in three parameters: regime (r.p.m.), stroke and diameter of the piston. This allows an entire rangeof flow between 350 cm^3/h and 2 cm^3/h, with maximum reproducibility. DELMAS (1979) and GARCIA HERNANDEZ (1981) give a detailes description of the experimental apparatus.

The kinetic curve is obtained by analysing the weathering solution, in such a way that the concentration and residence time (or reaction time) are related by means of the following equation:

$$t = \frac{V}{Q} = \int_0^C \frac{dC}{f(C)} \qquad \text{(equation 1)}$$

t = residence time.
V = pore volume.
Q = flux of weathering solution.
C = concentration.

The pore volume is related to the apparent bulk density and real density, according to the following equation:

$$V = V_{apparent} - V_{real} = \frac{m(1/\rho ap - 1/\rho real)}{Q} \qquad \text{(equation 2)}$$

Taking into account the values of m (mass of solid), apparent density (see table 1), the following equation is obtained that relates reaction times (hours) and flows (cm^3h):

$$t\,(h) = \frac{19.85}{Q\,(cm/h)} \qquad \text{(equation 3)}$$

The limit for dissolution is then obtained by projecting to $t \rightarrow \infty$ in a closed system and standard conditions (25°C and 1 atm.). This analysis allows estimation of the solubility and the solubility product constant (K_{sp}).

The crystals employed in this dissolution experiment in a closed system were submitted to a surface cleaning until the amorphous layer (Beilby layer) was removed and the stationary regime for a pure water flow of 70 cm^3/h (0.30 h of residence time) was established. They

Tab. 1:   PHYSICAL AND CHEMICAL CHARACTERISTIC

	SiO_2	FeO	Fe_2O_3	Al_2O_3	MnO	TiO_2	CaO	MgO	Na_2O	K_2O	Total
%	47.14	10.80	2.69	11.01	0.16	1.73	14.87	10.65	1.11	0.16	100.32 %

ø particles	=	0.05 – 2 mm
ρ apparent	=	1.965 $\pm$ 0.025 g/cm^3
ρ real	=	3.222 g/cm^3
m crystals	=	100.00 g/reactor

were later kept in contact with pure water, in a solid/solution ratio of 1/10, for three months, the stability of the final state of equilibrium being checked. A parallel experiment was also carried out under similar working conditions, with the exception that the mass of crystals used were not prewashed. This allowed study of the influence of the Beilby layer and of interface-solution equilibria on the values of K_{sp}.

The elements in solution were analyzed by atomic absorption spectrophotometry on a Perkin-Elmer spectrophotometer mod. 603, with the exception of silica, which was colorimetrically determines by the MORRISON & WILSON method (1963), with the absorbance of the reduced complex (blue), measured at 810 nm. on a UV–VIS Perkin-Elmer spectrophotometer mod. 550–A.

The rate of weathering solution was obtained by collecting the solution in polyethylene jars and weighing it in the interval of time employed.

The concentrations of the elements dissolved in the weathering solution are given in the table of results and are the mean values of 12 determinations carried out on alternative days for 4 weeks, with the system in a stationary working regime (see Figure 2).

In all cases, the standard deviation is $< 4\%$, both in the concentrations and flows of weathering solutions, selected as maximum limit of error for all the experimental points (pairs C,t).

The existence of surface defects, produced by the physico-mechanical treatment undergone by the crystals as their size is reduced, causes an amorphous zone at the surface in such a way that the product exhibits a peculiar behaviour that is well known in studies of dissolution. This results in an increase in the rate and amount of dissolution, which hinders the interpretation of mechanisms and results. It can be avoided by using the experimental device adopted, since it allows the evolution of the interface to be followed from the moment the device is set in motion until the stationary regime of dissolution is established. This evolution was studied from the monitoring of the variation of the concentration of $Si(OH)_4$, during the experiment and for each reactor.

The variation of the concentration of $Si(OH)_4$ with time shows an asymptotic decrease of exponential type with respect to time down to a value of concentration that corresponds to the stationary regime (see figure 2).

This value was attained for augite, after $\sim$ 4 months of uninterrupted monitoring.

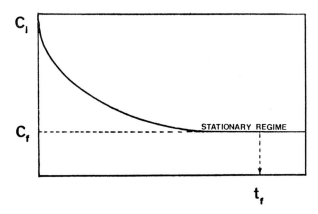

Fig. 2:   Variation of concentration since beginning of the experiment to the stationary regime.

The mineral selected was a basaltic augite extracted from the B/C horizon of a brown soil in Fasnia, South Tenerife. The monocrystals of augite selected were of medium size: between 2 and 4 cm. They were washed and submitted to ultrasound treatment, in order to remove the impurities adsorbed on the surface. They were later reduced in size by grinding in an agate mortar.

The sandlike mass thus obtained was freed of fine particles by sieving and particles of different density were removed by a countercurrent circulation of H_2O. The final product was of optimum mineralogical purity and crystallinity, as checked by X-ray diffraction and petrographic examination.

The physical and chemical data of the resulting augite are given in Table 1. The chemistry shows the mineral to be titaniferous, according to the nomenclature of YAGI & ONUMA (1967).

The stoichiometric formula is:

$$(Na_{0.032}K_{0.005}Mg_{0.496}Ca_{0.821}Fe^{+2}_{0.649}Fe^{+3}_{0.073}Mn_{0.010}Ti_{0.005}) (Si_{1.700}Al_{0.225}Ti_{0.075}) O_6$$

were Ti and Al are included as isomorphic substitutions of Si, following the suggestions of BARTH (1931) and SHKODZINSKIY (1968).

Tab. 2: EXPERIMENTAL VALUES

Reactor	SiO$_2$ 25°C t(h)	Cx10^6(M)	Cx10^6/t	pH	SiO$_2$ 32°C t(h)	Cx10^6(M)	SiO$_2$ 40°C t(h)	Cx10^6(M)
1	0.29	1.70	5.86	6.03	0.17	1.38	0.49	1.24
2	0.71	3.69	5.20	6.09	0.63	4.33	0.63	1.35
3	0.80	4.26	5.32	6.62	0.82	5.57	0.85	1.85
4	1.13	6.21	5.49	6.63	1.26	9.17	1.28	2.62
5	1.66	5.62	3.39	6.37	1.77	9.11	1.76	2.68
6	3.09	7.52	2.43	6.47	2.60	16.6	2.64	3.99
7	3.83	16.7	4.36	6.80	3.81	29.5	3.77	6.63
8	4.79	17.9	3.74	6.96	8.48	46.4	6.07	7.48
9	6.03	28.0	4.64	7.20				
10	31.51	91.5	2.90	7.35				

Reactor	MgO 25°C t(h)	Cx10^6(M)	CaO 25°C t(h)	Cx10^6(M)	K$_2$O 25°C t(h)	Cx10^6(M)	Na$_2$O 25°C t(h)	Cx10^6(M)
1	0.29	0	0.29	0	0.27	1.7	0.27	0
2	0.70	0.82	0.71	0	0.74	2.0	0.74	0
3	0.79	0.82	0.79	3.4	0.77	3.0	0.77	0
4	1.13	1.9	1.14	6.5	1.13	5.4	1.13	0
5	1.66	1.2	1.66	4.7	1.66	3.1	1.66	0
6	3.09	3.1	3.01	11	3.09	3.7	3.09	0
7	3.73	7.2	3.74	26	3.75	7.6	3.75	0
8	4.81	8.5	4.81	24	4.81	8.3	4.81	0
9	6.03	10	6.03	43	6.03	12	6.03	5.2
10	31.51	33	31.51	55	31.51	18	31.51	50

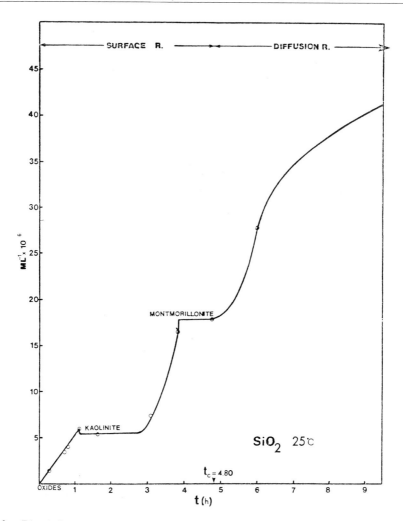

Fig. 3:   Dissolution curve for SiO₂ at 25°C.

3.   RESULTS AND DISCUSSION

Table 2 presents the mean concentration values of the weathering solution for all the species present together with the mean flows, pH and related kinetic parameters for the temperatures studied.

Figures 3, 4, 5 and 6 give the experimental values corresponding to silica for the three temperatures and for all the species dissolved at 25°C.

3.1.   KINETIC ANALYSIS

The inflections on the kinetic curves indicate a complex mechanism, the kinetic nature

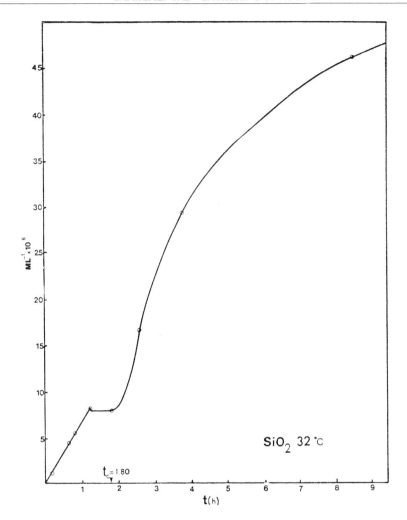

Fig. 4: Dissolution curve for SiO_2 at 32°C.

of which can be deduced from the study of the C/t quotient in the curve al 25°C (Table 2). It can be observed that the C/t quotient is constant for the first 4 reactors: that is, the rate is constant and independent of the diffusion regime (flow), in agreement with a control of the dissolution by surface reaction. The plateaus of concentration correspond to the nucleation and crystal growth of new phases that can be identified as kaolinite and montmorillonite, developed upon the oxides residual layer.

The sequence oxides → kaolinite → montmorillonite was described by GARCIA HERNANDEZ (1981) and DELMAS (1982), for the system cristobalite – Al^{+3} – H_2O, and implies a mechanism of epitaxial nucleation, where the surface of weathered mineral serves as a support for the process of nucleation and is associated with the dissolution controlled by the surface reaction.

The device does not permit the accumulation of sufficient neoformed product during

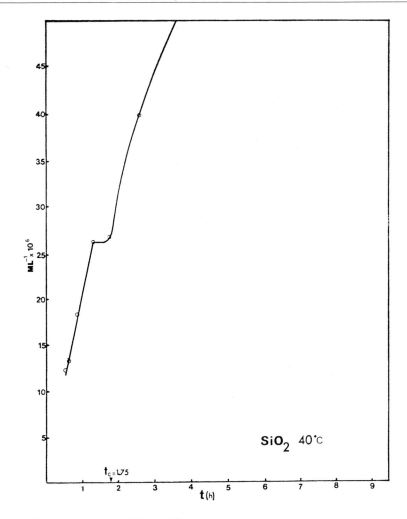

Fig. 5:   Dissolution curve for SiO_2 at 40°C.

the time of the experiment, to be detected by normal physical techniques (RX), but confirmation of the hypothesis is provided by TEJEDOR SALGUERO et al. (1978), who studied the clay fraction of the soil from which the mineral was extracted, where montmorillonite, kaolinite and iron oxides (goethite and hematites) predominated and where augite was predominant in the sand fraction.

Two clearly-defined zones can thus be distinguished in the kinetic curve: Surface reaction and Diffusion reaction, separated by a critical time (t_c) or time of exchange control in the transfer reaction. This t_c is present at the three temperatures studied, although it is observed that the t_c decreases as the temperature increases (Table 3), a fact interpreted as the result of a progressive saturation of the sites negavite charge asociated with the control of the dissolution by the surface reaction. This was reported by GARCIA HERNANDEZ (1981) as the normal behaviour of the crystalline silicates.

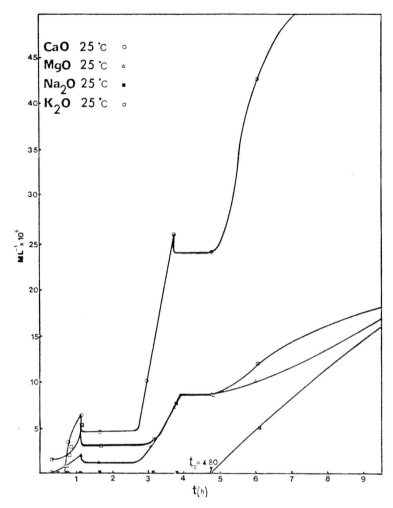

Fig. 6:   Dissolution curve for Cations at 25°C.

Tab. 3:   EVOLUTION OF t WITH T

T (°C)	t (h)
25	4.80
32	1.80
40	1.75

3.2. ANALYSIS OF THE CONGRUENCE OF THE DISSOLUTION

Two cases: Surface R. and Diffusion R. are distinguished in order to carry out this analysis.

Tab. 4:   PARAMETERS OF CONGRUENCE

	$b \times 10^6$	Surface react. (C = a + bt)		Diffusion react.	
		r_i (correl. coeff.)	b_i/b^*_i	C^*_i	C_i/C^*_i
Ca^{+2}	5.45	0.98	2.11	1.25×10^{-4}	0.38
Mg^{+2}	2.22	0.98	1.42	7.56×10^{-5}	0.99
Na^+	–	–	0	4.88×10^{-6}	94
K^+	4.20	0.97	267	7.63×10^{-7}	60
$Si(OH)_4$	5.35	1.00	1	2.08×10^{-4}	1 (refer.)

$$b^*_i = b_{Si(OH)_4}\left(\frac{v - i}{v\,Si(OH)_4}\right)$$

$$C^*_i = C_{Si(OH)_4}\left(\frac{v_i}{v\,Si(OH)_4}\right) \quad \text{equilibrium}$$

In surface reaction the parameter of congruence is deduced from the relationship between the slopes (b_i) of the experimental straight lines and those obtained (b^*_i) by multiplying the slope of the straight line of the SiO_2 solution, taken as reference, by the quotient of the stoichiometric coefficient of the element considered (v_i) and that of silica ($v_i\,Si(OH)_4$ = 1.700).

In diffusion reaction, the comparison is carried out in an analogous manner by considering the limiting value of the solubility for each species dissolved (Table 6), and the relationship of coefficients (v_i) in the equation of equilibrium (Equation 7) (see Table 4).

In Table 4, we can observe for the alkaline-earth cations, which are in excess in surface reaction, a tendency to stoichiometry in difussion reaction for the Mg^{+2}, while the Ca^{+2} presents a deficit two and half times greater with regard to the stoichiometric relationship. Both decreases are interpreted as due to occlusion during the steps of neoformation, as deduced from Figure 6 (probably as carbonate paragenesis episode).

The dissolution is particularly selective for the alcaline cations both in surface reaction and diffusion reaction. The K^+ presents a similar behaviour during all reaction. The Na^+ displays a strange behaviour; while in surface reaction it is occluded in the transfer interface, in diffusion reaction it is dissolved spectacularly to a ratio 94 times greater than the stoichiometric when the solution is in equilibrium.

It can be observed that the non-congruence for all the cations dissolved increases in the sense:

$$Mg^{+2} < Ca^{+2} < K^+ < Na^+$$

while the ratio charge/ionic radius decreases as follows:

$$Mg^{+2}\,(3.08) > Ca^{+2}\,(2.02) > Na^+\,(1.11) > K^+\,(0.7)$$

The tendency to non-stoichiometry is seen to increase as the ratio charge/ionic radius decreases, Na^+ being an exception that presents a specific behaviour.

This unique behaviour of the Na^+ is interpreted in surface reaction as an occlusion of the Na^+ in the residual layer, as pointed out by ILER (1979), who applied "Site-Binding" theory;

in diffusion reation, the Nernst layer of transfer facilitates the diffusion of the Na^+ ions towards the solution, in a manner similar to that proposed by HOPPE (1941) for the dissolution of alkaline elements, from a vitreous porous layer.

This tendency of the Na^+ is a result of the transition of the electrical nature of the interface: from a negative interface (surface reaction) to a neutral interface (diffusion reaction), in a manner similar to that described by GARCIA HERNANDEZ (1981) as normal behaviour of crystalline silicates that are dissolved by mixed mechanism: surface-diffusion.

Similar results, with regard to the selectivity of the dissolution of Na^+ and K^+ in augite, have been obtained by KELLER et al. (1963), although the factor of non-stoichiometry found by these authors varies between 14 and 48.

PACES (1973) explains the selectivity of the Na^+ in the dissolution of albite, as due to a $Na^+ \leftrightarrow \subset^+$ exchange through a diffuse layer, according to first order kinetics. In our case, we propose the following model, expressed in equation 4.

$$Si-O-Na + H^+ \xrightarrow[\text{Surface r.}]{\text{Diffusion r.}} Si-OH + Na^+ \quad \text{(equation 4)}$$

The equation expresses the tendency of the reaction. In surface reaction, the Na^+ is associated with the negative sites produced by the dissociation of the $\equiv Si-OH$ and is retained in the residual layer. In diffusion reaction, the tendency is inverted and the $Na^+ - H^+$ exchange takes place, in agreement with the slight alkalinization of the solution observed from t (reactors 9 and 10, Table 2).

3.3. CALCULATION OF THE ACTIVATION ENERGY ($\triangle G^{\neq}$).

The $\triangle G^{\neq}$ is obtained by applying the Arrhenius law, according to the method proposed by AMMOU–CHOKROUM et al. (1977) for chalcopyrite, where the logarithmus of the slopes of the tangents from the origin ($t = 0$) to the curves of dissolution, are plotted against $1/T$.

In our case, the slopes of the straight lines of the dissolution of silica at the three temperatures are employed (Table 5).

The logarithmus of the slopes against $1/T$ are plotted in Figure 4.

The non-optimus correlation coefficient (0.96) of the straight line thus adjusted: $\ln K = 8.49 - 6.1 \times 10^{-3}/T$, could be interpreted as due to a deviation from the Arrhenius law, whereby the variation of the rate constant with T would not simply be a decreasing exponen-

Tab. 5: EQUATIONS OF THE STRAIGHT LINES OF LINEAR REGRESSION FOR SILICA WITH REGARD TO T

$$C(M) = a + b\ t(h)$$

T°C	a	b	Correlation Coeff.
25	4.45×10^{-8}	5.35×10^{-6}	0.99_7
32	-1.54×10^{-8}	7.12×10^{-6}	0.99_7
40	2.91×10^{-6}	1.81^{-5}	0.99_5

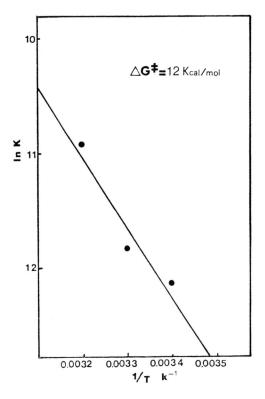

Fig. 7:   Plotting of the Arrhenius equation.

tial functon (equation 5), but rather the result of an exponential by a potential (equation 6), as pointedout by FROST & PEARSON (1953).

$$K = A e^{-\triangle G^{\ne} /RT}$$ (equation 5)

$$K = A' T^m e^{- \triangle G^{\ne} /RT}$$ (equation 6)

The solution of equation 6 by successive approximations, in the case under study, converges rapidly to the same value of $\triangle G^{\ne}$ obtained (12.12 Kcal/mol) from the Arrhenius equation (equation 5). These differences are throught to be due, therefore, to changes in the mechanism of dissolution when T increases, and not to deviation to Arrhenius law.

The value of $\triangle G^{\ne}$ obtained (12.12 Kcal/mol) for the activation energy of the dissolution of augite, can readily be compared with those obtained by BAILEY (1974) for wollastonite (13 Kcal/mol) and by GRANDSTAFF (1977) for bronzite (10.5 Kcal/mol).

3.4.   THERMODYNAMIC PARAMETERS

The values of solubility at 25°C and 1 atm. for the mass of crystals submitted to surface washing are presented in Table 6 (experiment n. 1), compared with those obtained under the

Tab. 6: SOLUBILITY VALUES

	Experiment no. 1 (M)	Experiment no. 2 (M)
$Si(OH)_4$	2.08×10^{-4}	8.46×10^{-4}
Ca^{+2}	4.8×10^{-5}	7.5×10^{-6}
Mg^{+2}	7.5×10^{-5}	5.4×10^{-6}
Na^+	4.5×10^{-4}	1.2×10^{-3}
K^+	4.6×10^{-5}	1.4×10^{-3}
OH^-	2.8×10^{-6}	3.6×10^{-6}

same conditions for the untreated mineral (experiment n. 2).

It can be observed that the values of the concentrations in experiment n. 1 are higher than those of experiment n. 2, with the exception of Ca^{+2} and Mg^{+2} for which the tendency is the reverse. This is considered to be due to occlusion in the amorphous layer.

In any case, a considerable difference is observed in the quantitative amount of solution in the two cases, which is in agreement with the explanation of the surface characteristics of the crystals employed in the dissolution experiments.

The equilibrium equation (equation 7) can be deduced by taking into account the stoichiometric formula and the neoformed phases, approximation by their ideal formulae:

$$(Na_{0.032}K_{0.005}Mg_{0.496}Ca_{0.821}Fe^{+2}_{0.649}Fe^{+3}_{0.073}Mn_{0.010}Ti_{0.005}) \ (Si_{1.700}Al_{0.225}Ti_{0.075})O_6 \ (s)$$

$+ \ 5.325 \ H_2O + 0.163 \ O_2 \rightleftharpoons 0.722 \ Fe(OH)_3(s) + 0.080 \ TiO_2 \ (s) + 0.010 \ Mn(OH)_2 \ (s) +$ $0.562 \ Al_2Si_2O_5(OH)_4 \ (s) + 0.056 \ Al_2Si_4O_{10}(OH)_2 \ (s) + 0.005 \ K^+ + 0.032 \ Na^+ + 0.496$ $Mg^{+2} + 1.364 \ Si(OH)_4 + 0.821 \ Ca^{+2} + 2.671 \ OH^-$

$$K_{sp} = \pi_i a_i^{v_i} = 2.70 \times 10^{-26} \qquad \text{(equation 7)}$$

The calculation of the activity coefficients of the ionic species was carried out on the bases of the simplified Debye-Hückel equation (equation 8):

$$\log \gamma_i = - A Z_i^2 \ \mu^{1/2} \qquad A = 0.509 \ (H_2O \ a \ 25°C), \ \text{(equation 8)}$$

The ionic strength (μ) was obtained by applying the equation by GRIFFIN & JURINAK (1973) (equation 9), obtained for natural waters:

$$\mu = 0.013 \ E.C. \qquad \text{(equation 9)}$$

where E.C. is the conductivity in $mmhos/cm^2$.

Thus, $\mu = 1.098 \times 10^{-3}$ M and the following activity coefficients are obtained (Table 7). These values almost equal one, and the activities were considered, therefore, equal to con-

Tab. 7: ACTIVITY COEFFICIENTS OF DISSOLVED SPECIES

Ions	γ_i
M^{+2}	0.99_7
M^+	0.99_9

centrations.

Application of the equilibrium equation (equation 10) that relates the variation of free energy ($\triangle G^0_{diss}$), to the thermodynamic constant of equilibrium (K_{sp}) allows the $\triangle G^0 f$ of augite to be calculated (equation 11),

$$\triangle G^0_{diss} = \sum \triangle G^0_f \text{ (reagents)} = -RT \ln K_{sp} \qquad \text{(equation 10)}$$

$$\triangle G^0_f \text{ (augite)} = -621.53 + 1.364 \log K_{sp} \qquad \text{(equation 11)}$$

leading to a value of $\triangle G^0 f$ (augite) of –676.5 Kcal/mol, clearly similar to that obtained by HUANG & KELLER (1972) for another augite (–679.4 Kcal/mol).

4. CONCLUSIONS

– The dissolution of augite is controlled by a mixed mechanism: surface-diffusion; it is non-congruent in nature and selective as regards the dissolution of alkaline elements, especially Na^+.

– The value of activation energy ($\triangle G^+$) of the reaction of dissolution of augite is 12 Kcal/mol.

– The value of the $K_{sp} = 2.70 \times 10^{-26}$, and that of the free energy of formation of augite is $\triangle G^0_f = -676.5$ Kcal/mol.

– The existence of mixed mechanisms: Surface reaction–Diffusion reaction is revealed as a generalized tendency in the dissolution of crystalline silicates.

BIBLIOGRAPHY

AMMOU–CHOKROUM, M., CAMBAZOGLU, M. & STEIMETS, D. (1977): Oxydation menagee de la chalcopyrite en solution acide: Analyse cinetique des reactions. I. Modeles Chimiques. II. Modeles diffusionels. Bull. Soc. fran. Miner. Crist., **100**, 149-177.

BAILEY, A. (1971): Comparison of Low-temperature with High-temperatures diffusion of sodium in albite. Geochim. Cosmichim. Acta, **35**, 1073-1081.

BAILEY, A. (1974): Effects of temperature on the reaction of silicates with aqueous solutions in the low temperatures range. Proc. Int. Symposium on water-rocks interactions (Praga), 375-380.

BARTH, T.F.W. (1931): Pyroxen von Hiva Oa, Marquesas-Inseln und die Formel titarhaltiger Augite. Neues Jahrb. Min., Abt. A., **64**, 217-224.

DELMAS, A.B., PEDRO, G. (1971): Sur l'alteration experimentale de l'olivine par lessivage a l'eau et la misse en evidence de trois grands domaines d'evolution geochimique. Acad. Sci. Paris, **273**, 1543-1546.

DELMAS, A.B. (1979): Etude experimentale du phenomene de dissolution des sels simples et des silicates. Approche cinetique. Thèse. Ser. Publications INRA (Versailles).

DELMAS, A.B., GARCIA HERNANDEZ, J.E. & PEDRO, G. (1982): Discussion sur les conditions et les mecanismes de formation du quartz a 25° C en milieu ouvert. Analyse cinetique. Sci. Geol. Bull., **35**, 81-91.

ERNST, W.G. & CALVERT, S.E. (1969): An experimental study of the recrystallization of porcelanite and its bearing on the origin of some bedded cherts. Am. J. Sci. Schairer, **267**, 114-133.

FROST, A.A. & PEARSON, R.G. (1953): Kinetics and Mechanism. John Wiley, New York.

GARCIA HERNANDEZ, J.E. (1981): Interpretation cinetique de la geochimie d'alteration de la silice a basse temperature 25°C. Thèse. Serv. Publications INRA (Versailles).

GARCIA HERNANDEZ, J.E. (1982): Modelos dinámicos en el estudio de la génesis del suelo. An. Edaf. Agrobiolog., **41**, 1097-1103.

GRANDSTAFF, D.E. (1977): Some kinetics bronzite orthopyroxene dissolution. Geochim. Cosmo-

chim. Acta, **41**, 1097-1103.
GRIFFIN, R.A. & JURINAK, J.J. (1973): Estimation of activity coefficients from the electrical conductivity of natural aquatic systems and soils extracts. Soil Sci., **116**, 26-30.
HELGESON, H.C. (1971): Kinetics of mass transfer among silicates and aqueous solutions. Geochim. Cosmochim. Acta, **35**, 421-469.
HOPPE, H.J. (1941): Chem. D. Erde, **13**, 484. In: CORRENS, C.W. (1961): The experimental chemical weathering of silicates. Clay Miner. Bull., **4**, 249-265.
HUANG, W.H. & KELLER, W.D. (1970): Dissolution of rock-forming silicate minerals in organic acids: Simulated first-stage weathering of fresh mineral surfaces. Am. Mineralogist, **33**, 2076-2094.
HUANG, W.H. & KELLER, W.D. (1972): Standard free energies of formation calculated from dissolution data using specific mineral analyses. Am. Mineralogist, **57**, 1152-1162.
ILER, R.K. (1979): The chemistry of Silica. Wiley Interscience, New York.
KELLER, W.D., BALGORD, W.D., REESMAN, A.L. (1963): Dissolved products of artificially pulverized silicate minerals and rocks: Part I. J. of Sedimentary Petrolog., **33**, 191-204.
LUCE, R.W., BARTLETT, R.W. & PARKS, G.A. (1972): Dissolution kinetics of magnesium silicates and aqueous solutions. Geochim. Cosmochim. Acta, **36**, 35-50.
MORRISON, I.R., WILSON, A.L. (1963): The adsorptiometric determination of silicon in water. Analyst, **88**, 100-104.
PACES, T. (1973): Steady state kinetics and equilibrium between ground water and granitic rocks. Geochim. Cosmochim. Acta, **37**, 2641-2663.
SHKODZINSKIY, V.S. (1968): The position of titanium in clinopyroxene. Dokl. Acad. Sci. U.S.S.R., Earth Sci. Sect., **182**, 142-145.
TEJEDOR SALGUERO, M.L., QUANTIN, P. & FENANDEZ CALDAS, E. (1978): Climatoséquence de la région méridionale de l'île de Ténerife (Iles Canaries). 2. partie: Caractéristiques minéralogiques: interprétation et classification. Cah. ORSTOM Ser. Pedol., **16**, 83-106.
YAGUI, K. & ONUMA, K. (1967): the join Ca Mg Si2O6 – Ca Ti Al2O6 and its bearing on the titanaugites. J. Fac. Sci. Hokkaido Univ., serv. 4, **13**, 463-483.

Address of authors:
González Bonmati, J., Vera Gómez, M.P., García Hernández, J.E.
Departamento de Edafología, Universidad de La Laguna,
Tenerife. Spain.

5

E. Fernandez Caldas & Dan H. Yaalon (Eds):
VOLCANIC SOILS
CATENA SUPPLEMENT 7, Braunschweig 1985

HIGH–CONCENTRATION LEVELS OF HEAVY MINERALS IN TWO VOLCANIC SOILS FROM COLOMBIA: A POSSIBLE PALEOENVIRONMENTAL INTERPRETATION

P.A. Riezebos, Amsterdam

SUMMARY

Detailed petrographic descriptions of the horizons recognised in two volcanic soils from Colombia (S. America) reveal at some levels increased heavy-mineral abundances. In one soil this increase coincides with the B_2 and A_{12} horizons and in the other one with the BC and B_2 horizons. These zones of enrichment have been interpreted as a sort of aeolian residual accumulation and are associated with drier and colder climatic fluctuations during a former development phase of the soils. This phenomenon might be of use as an additional aid in paleoenvironmental reconstructions.

ZUSAMMENFASSUNG

Detaillierte petrographische Beschreibungen von den erkannten Bodenhorizonten in zwei vulkanischen Böden von Colombia (S. America) ergeben erhöhte Schwermineralvorkommen auf bestimmten Niveaus, in einem Boden zusammenfallend mit dem B_2 und A_{12} Horizont, und im anderen mit dem BC und B_2 Horizont. Wenn diese Niveaus an der Oberfläche beobachtet werden, dann können diese Erscheinungen von Nutzen sein als zusätzliches Hilfsmittel in paläo-umweltlichen Rekonstruktionen. In diesem Artikel wurden diese Anreicherungszonen erklärt als eine Art von äolischem Residualkonzentrat.

1. INTRODUCTION

In the soil mantle heavy minerals occur as residual and/or as supplied components. Compositional differences of the heavy-mineral fractions separated from sequences of soil horizons have been reported frequently, and were used successfully in investigations of various nature (e.g. HASEMANN & MARSHALL 1945, and FRIED 1983).

As far as the present author is aware, little attention has been given to variations in the concentration of detrital heavy-mineral separates as a function of soil depth. ZIMMERLE (1973), giving a tentative classification of heavy-mineral concentrations as to their depositional environment, estimates their relative abundance in non-fluviatile, continental environments as rare. As soils are frequently of a rather ephemeral nature (geologically spoken), it is not surprising that the soil environment was not enclosed in ZIMMERLE's classification. Therefore, few soils are likely to prove economic sources of useful detrital heavy minerals. Possibly such soils might only be found on cassiterite-bearing rocks, for instance.

In this paper, the phenomenon of varying heavy-mineral abundances as a function of soil depth will be shown and discussed in two volcanic soils from Colombia (S. America). The work forms part of a research project on the Quaternary of Colombia of Dr. T. van der Hammen (Palynological Department of the Hugo de Vries Laboratory, University of Amsterdam).

ISSN 0722–0723 / ISBN 3–923381–06–9

Photo 1:   The Facatativa soil section.

Photo 2:   A soil section rather similar and very near to the investigated Guasca section.

2. SITE AND GENERAL SETTING OF THE PROFILES

The soils are found on the Sabana de Bogotá, a large upland plain in the Cordillera Oriental. The main uplift of this mountain range occurred between 4.5 and 2.5 million years ago (VAN DER HAMMEN et al. 1973). During the Pleistocene this plain was covered by a large lake. Within the Pliocene-Quaternary sedimentary infill of this lake three formations are distinguished: the Tilatá Formation (Pliocene), the Subachoque Formation (Pliocene) and the Sabana Formation (Pleistocene).

The first soil section (Photo 1) is situated along the road between Facatativa und El Rosal (Fig. 1). It is found at an altitude of about 2615 m on a southeast facing slope. According to the Geological Map of Colombia, Sheet K–10 Villeta, the Q1b unit forms the surface of this site. The second soil section (Photo 2) occurs along the road between Guasca and Sueva at an elevation of about 2805 m on a westward facing slope (Fig. 1). According to Sheet K–11 Zipaguira of the Geological Map, at this spot the Qal unit occurs. The Qal and Q1b units are generally described as fine-grained fluvio-lacustrine deposits, and both belong to the Sabana Formation. They are, however, covered by, and highly mixed with, volcanic material giving rise – as field evidence shows – to the development of volcanic soils.

Deposition of pyroclastic material in the region of the Sabana de Bogotá has been a frequent event. The ashes are crystal-vitric of nature and have been laid down several times in the area of Sabana de Bogotá during the last 45.000 years (RIEZEBOS 1978). The Quaternary volcanism is assumed to have occurred in the Cordillera Central, so the volcanic centres found in this mountain range at a distance of about 150 km have very probably produced the ashes.

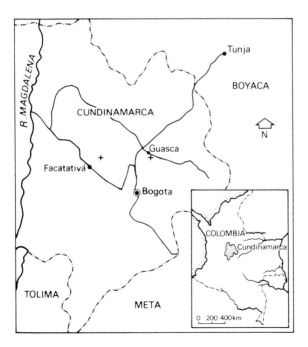

Fig. 1:   Map of Cundinamarca (Colombia) showing the sites (+) of the soil profiles.

The present climate of the Sabana de Bogotá shows two wet (April–May and October––November) and two dry seasons. The annual rainfall values are varying between 700 and 900 mm. In places, however, these values may be lower or higher (VAN DER HAMMEN & GONZALEZ 1963). BOTERO (1972) reports that east of the Sabana de Bogotá, in the adjoining, topographically higher area of the Eastern Cordillera, the annual precipitation varies between 1800 and 2200 mm, falling mainly in the period May-August. The average annual temperature on the Sabana de Bogotá is reported to range between 13 and 15°C, showing otherwise a close relationship with altitude. This relationship amounts to ⅔°C/100 m according to VAN DER HAMMEN & GONZALEZ (1963). Calculations show the present annual temperature at the Guasca soil site to be 11-10°C and at the Facatative soil site 12-11°C.

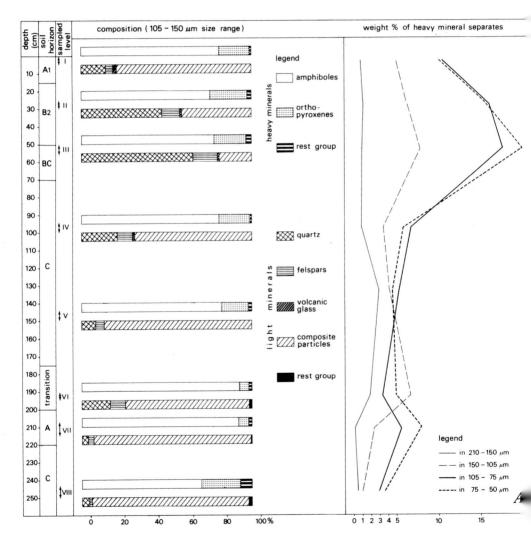

Fig. 2: Petrographical data of the Guasca (A) and the Facatativa (B) soil sections. The rest group in the light-mineral fractions consists mainly of polymineral- and phyllosilicate aggregates and rock fragments.

3. SAMPLE TREATMENT AND METHODS

From the Facatativa profile, which has been studied earlier by JUNGERIUS (1975, 1976), only bulk samples representing the distinguished soil horizons were available. From the Guasca profile (RIEZEBOS & LUSTENHOUWER 1983) 5-10 cm thick samples were collected at regular intervals. After suspension in demineralised water, the material was gently pulverised, treated with H_2O_2 (30%), repeatedly washed, and after drying the different sand-sized fractions ($> 32 \mu$) were isolated by sieving. The reasons for omitting the usual treatment with acid or alkaline digestion has been given elsewhere (RIEZEBOS & LUSTENHOUWER 1983). Heavy minerals were separated from four sand fractions by using bromoform (s.g. 2.89) and their weight percentages were recorded. The composition of the heavy-mineral separates and the remaining light-mineral fractions were established using the petrological microscope by analysing 300 translucent particles.

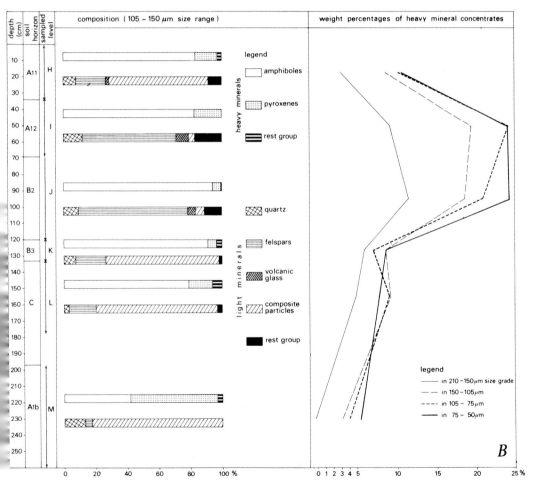

In the heavy-mineral fractions this group comprises epidote, garnet, mica, anatase, zircon, tourmaline, brookite and sphene.

Tab. 1: A. HEAVY MINERAL COMPOSITIONS IN THE GUASCA SECTION IN TERMS OF AMPHIBOLES, ORTHOPYROXENES AND A REST GROUP COMPRISING MICA, GARNET, EPIDOTE, ANATASE, ZIRCON AND TOURMALINE. VALUES WITH ASTERISKS CONTAIN LESS THAN 1% CLINOPYROXENES.
B. HEAVY–MINERAL COMPOSITIONS IN THE FACATATIA SECTION IN TERMS OF AMPHIBOLES, PYROXENES AND A REST GROUP COMPRISING EPIDOTE, MICA, GARNET, TOURMALINE, ZIRCON, BROOKITE AND SPHENE.

A

Depth (cm)	210-150 μ			150-105 μ			105-75 μ			75-50 μ		
	amphiboles	ortho-pyroxenes	rest group	amphiboles	ortho pyroxenes	rest group	amphiboles	ortho-pyroxenes	rest group	amphiboles	ortho-pyroxenes	rest group
0-5	76.7	20.7	2.6	80.7	18.0*	1.3	87.4	10.7	1.9	79.4	16.0	4.6
25-30	75.3	23.0*	1.7	76.0	20.7	3.3	77.7	19.7*	2.6	76.3	19.7	4.0
50-56	77.1	21.0*	1.9	78.4	18.3*	3.3	74.3	23.0*	2.7	71.5	25.6*	2.9
94-100	82.1	14.3	3.6	85.6	12.7	1.7	82.6	13.7	3.7	83.4	11.7	4.9
144-150	85.2	10.4	4.4	83.4	14.3	2.3	80.6	13.3*	6.1	83.6	9.4*	7.0
190-195	93.7	5.3	1.0	92.7	5.0	2.3	93.6	3.7	2.7	88.5	6.3*	5.2
207-215	92.8	5.0	2.2	92.0	6.0	2.0	94.4	3.6*	2.0	90.6	4.7	4.7
243-250	66.4	26.6	7.0	69.0	24.3	6.7	74.3	18.7	7.0	78.3	14.0*	7.7

B

Depth (cm)	210-150 μ			150-105 μ			105-75 μ			75-50 μ		
	amphiboles	pyroxenes	rest group	amphiboles	pyroxenes	rest group	amphiboles	pyroxenes	rest group	amphiboles	pyroxenes	rest group
0-33	75.3	21.4	3.3	83.4	14.0	2.6	82.4	15.3	2.3	75.4	21.0	3.6
33-70	87.7	10.3	2.0	82.3	17.7	–	81.3	17.4	0.3	76.7	23.0	0.3
70-120	94.0	4.7	1.3	93.3	6.0	0.7	92.3	6.7	1.0	89.3	9.4	1.3
120-133	94.1	3.0	2.9	91.0	5.7	3.3	87.4	8.3	4.3	85.7	10.3	4.0
133-178	74.7	22.0	3.3	79.0	16.0	5.0	81.0	14.3	4.7	80.8	13.6	5.6
198-260	28.3	70.7	1.0	43.0	54.0	3.0	53.6	42.7	3.7	65.0	27.7	7.3

4. RESULTS

The main results have been summarized in Figure 2. At each sampled depth, the upper bar represents the heavy-mineral composition and the lower one that of the light minerals. Only the mineral spectra of the 105-150 μ size grades have been depicted. Similar ratios have been observed also in the other sand-sized fractions (Table 1).

In both profiles the heavy-mineral spectra appear to be dominated by amphiboles. Pyroxenes form the most important accessory component. In the Guasca profile they consist almost exclusively of orthopyroxenes. In the Facatativa profile, however, also clinopyroxenes are found, although in minor amounts (Table 1).

A distinction of the sampled horizons on basis of their heavy-mineral assemblages is hardly possible. Only the buried A1 horizon in the Facatativa profile can plainly be discriminated from the overlying soil with the aid of the heavy-mineral composition. However, on a more close inspection of the data in Table 1, both soil profiles appear to have levels with comparatively lower pyroxene contents in all of the size-grades analysed. In the Guasca soil this concerns the depths 190-195 cm and 207-215 cm, in the Facatativa soil the depth 70-133 cm. These consistently lower pyroxene abundances are due to a reduction of the orthocombic component exclusively. It is very tempting to see this in both profiles as a common characteristic, reflecting a similar source of origin, and implying simultaneity of deposition of the materials at the depths mentioned.

The light-mineral compositions show more variation. In the Guasca profile most mineral spectra are dominated by composite particles which are considered pedorelics cemented by allophanic material (RIEZEBOS & LUSTENHOUWER 1983) except at two depths corresponding with the B_2 and BC horizons (levels II and III). Here quartz constitutes the major component with composite particles, felspar, and a few percents of volcanic glass as minor constituents. A similar picture is found in the Facatativa profile. At all depths the composite particles dominate the mineral spectra, except in the A_{12} and B_2 horizons (levels I and J) where felspar prevails followed by quartz, volcanic glass, and composite particles as accessories.

Curves representing the heavy-mineral content in the four size grades as a function of soil depth exhibit two conspicuous characteristics. Firstly, this soil material contains a higher heavy-mineral abundance than soil materials derived from more mature sediments. The second remarkable feature is that in both profiles certain levels are discernible at which the heavy-mineral content in almost all size-fractions shows a sudden increase. In the Guasca profile this rise corresponds with the BC and B_2, in the Facatativa profile with the B_2 and A_{12} horizons. Finally, attention is focussed on the fact that these levels of enrichment in both profiles are associated with light-mineral spectra which deviate from the overlying and underlying soil material.

5. DISCUSSION

Abundances of heavy minerals found in airborne pyroclastic deposits will not simply reflect the initial ratio between heavy (s.g. > 2.89) and light (s.g. < 2.89) components in the original products ejected from the erupting vent. Heavy-mineral contents in volcanic ashes and in their derivative deposits in the landscape are controlled by numerous variables.

Long ago "eolian differentiation" has been recognised (LARSSON 1937, VAN BEMMELEN 1949), holding that the silica content of an ash increases with increasing distance from the vent. In terms of heavy-mineral contents this means that in the proximal portions of one single ash fall the heavy-mineral content is expected to be higher than in its distal areas. Superimposed upon this effect a vertical differentiation according to size, shape and density may occur in tephra accumulated for the first time in the earth's surface. This may lead to vertical variations in abundances. Recently this was shown in the downwind tephra lobe of Mount St. Helen's eruption in 1980 (JUVIGNE & SHIPLEY 1983). In desert environment this eolian differentiation has likewise been noticed (SIDERENKO 1956).

Post-depositional changes, including both differential dissolution processes and different types of reworking on the ground surface will also affect the heavy-mineral abundances. If the tephra contain abundant unstable components e.g. shard fragments, in particular the differential dissolution would contribute to a slow but continuous relative increase of the heavy mineral content. The comparatively high weight percentages of the heavy concentrates in the present soil materials may thus have been produced by a variety of causes.

However, the remarkable rise of the heavy-mineral contents at the levels of the BC and B_2 horizons in the Guasca soil and of the B_2 and A_{12} horizons in the Facatativa soil represents a twice- to fourfold increase compared with the under- and overlying soil material. Because of this, the increase is probably not due to differential alteration in the soils, but to some drastic change in environmental conditions during the accumulation of the parent mterial.

Residual erichment by invoking an intensified dissolution of light-mineral constituents at these levels in the present soils has to be rejected as an explanation for two reasons. Firstly, in neither of the profiles do these enrichments coincide with zones recognized as eluvial soil horizons. Secondly, it is difficult to conceive how vertical or lateral water movements acting in materials as described in the C horizons of the soils, would produce mineral compositions as found to be associated with the conspicuously enriched levels.

Another possibility is that the levels enriched in heavy minerals represent "exotic" materials. An exclusive supply of this exotic material from nearby terrains upslope by creep or surface wash, however, seems unlikely. In bot soil sections the materials below and above the zones rich in heavy minerals (levels I and IV in the Guasca soil; levels H and K in the Facatativa soil) are petrographically rather similar. So the levels II and III (Guasca) and I and J (Facatativa) derive in any case partly from other, relatively far-off sources (Fig. 2).

From an earlier study (RIEZEBOS & LUSTENHOUWER 1983, 206) it appeared that the Guasca soil section displays a fourfold-stratified arrangement on the basis of the SiO_2/Al_2O_3 molar ratios established in the allophanic cement of the composite particles. Significant different values of this ratio were a.o. found between levels VI and V corresponding more or less with the boundary between the buried soil and the overlying C material of the surface soil. Also betwen the levels IV and III in the surface soil, being the transition from the C horizon to the BC horizon, such a difference was established. This indicates a change in nature between the material from level IV and that from level III, in particular because it is accompanied with an evident change in the light-mineral composition. In a single section it is hardly possible to find out whether that change occurred slowly in the course of a continuous accumulation, or that it took place rather suddenly upon a distinct erosion or deposition surface. Anyway, analogue reasoning suggests that a similar event occurred in the Facatativa soil between the B_3 and B_2 horizons.

In view of the compositional differences of the enriched levels in both soil sections, a supply of the added material from one single source area seems unlikely. In the Facatativa soil, the levels rich in heavy minerals might represent a rather pure pyroclastic ash, but this is certainly not the case in the Guasca soil, although here these levels contain evident volcanoclastic components. As the levels marked by high heavy-mineral contents neither seem to be derived totally from localities adjacent to the soil sites, a supplementary supply primarily from the air seems the most obvious conclusion.

The presence of airborne accretionary zones of different mineralogical composition in two surface soils at a distance of about 50 km, implies the prevalence of specific environmental conditions. Conditions with a climate drier than in the preceding period, with hardly any precipitation or distinct dry periods, and a much more open vegetation or no vegetation at all seems most likely. Such a landscape, covered by a mantle of loose, dry material totally or partly consisting of volcanic clastic components, also strongly suggests the nature of the enrichment mechanism.

An eolian supply of fine-grained loose material under dry conditions, followed by a continuously repeated reworking by the wind, must lead to a gradual removal of aerodynamically lighter particles. Which factors primarily control this removal and to which degree each factor has affected the ultimate result, is impossible to assess. In addition to the factors discussed above, size, shape and density may have been playing a role, each time again in the repeated depositional processes according to the hydraulic equivalence concept. But also the mode of transport – suspension, saltation, rolling –, surface roughness, repeated entrainment and displacements (GRIGG & RATHBUN 1969, SLINGERLAND 1977, SALLENGER 1979, STEIDTMANN 1982, etc.) may have contributed to an incrase of the heavy-mineral particles.

Within a given range of particle-size distribution, particle shape, bed configuration, flow characteristics etc., it might be expected that the dislodgement of high-density particles by wind from a bed surface will occur less easier, and therefore less frequently than the removal of low-density particles. This could result in a residual enrichment of heavy minerals in the bed surface. Further, when airborne loose deposits are subjected to a re-entrainment by wind, there is a continuous exchange of particles between the surface (the bed), the bedload (rolling and saltating population), and the suspended load. When a steady flow is assumed, after a certain time a steady state might be established, holding that the weights of the suspended load and the bedload remain more or less constant. But, as the constituting particles of these loads are continuously exchanged, it is thinkable that during the process of exchange a shift in the ratio heavy-light particles takes place, especially in the population of suspended particles. Material travelling out of the area – and mostly this concerns the suspended load – will show a relative impoverishment in heavy minerals.

So, if in a certain area with variable (or opposite) wind directions, the same material is continuously subjected to eolian reworking without being completely removed from that area, and this reworking occurs over a sufficiently long period of time, it might gradually lead to a relative heavy-mineral increase. In particular this will take place within those particle-sizes which have constantly been liable to a back and forward transport in suspension.

Summarizing, the higher heavy-mineral abundances assessed in some levels of these soils are considered as a feature, which is commonly found more fully developed in the so-called lag concentrations. In such concentrates a residual enrichment of some components occurred due to their larger size, their higher density, or their deviating shape. In the present case it is thought that in particular eolian winnowing during a continuously repeated migration has been the main mechanism in producing this phenomenon.

5.1. SPECULATIONS ON HE AGE OF THIS PHENOMENON

As already noticed, the occurrence of the enriched levels in two surface soils at a distance

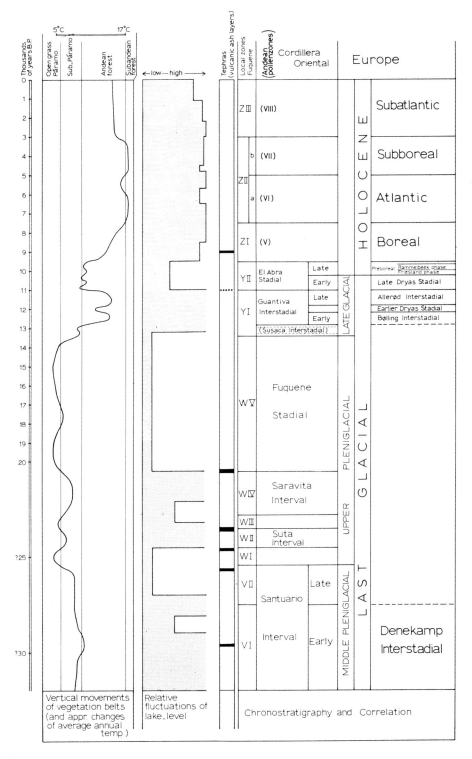

Fig. 3: Diagram showing climatic variations as inferred from palynological studies in the Fuquene area (according to VAN GEEL & VAN DER HAMMEN 1973).

of about 50 km suggests this phenomenon to be of regional significance, reflecting a comparatively dry and vegetation-poor period in which displacement and differentiation of surface material by wind must have been predominant. Two interesting questions arise:
 a) in what part of the Quaternary history did it occur, and
 b) did these enrichments take place in the same episode or in two different ones.

In Fig. 3 a diagram according to VAN GEEL & VAN DER HAMMEN (1973) has been reproduced. It illustrates the variations in vegetation zones and lake levels as has been deduced from palynological investigations in the Fuquene area about 80 km north of Guasca (Fig. 1). These climatic variations have been broadly confirmed by later palynological work on the Sabana de Bogotá (CORREAL URREGO & VAN DER HAMMEN 1977, VAN DER HAMMEN 1978, etc.). The diagram shows the El Abra Stadial (ca. 11000-9500 BP) and the Fuquene Stadial (ca. 21000-14000 BP) as the most recent periods to be considered.

In order to answer the questions posed, only petrographic characteristics are available in the soil profiles. So, attention is focussed on the fact that in both sections the buried A horizon has developed in volcanic material that 1) contains (ortho)pyroxenes, 2) is devoid of shards, and 3) is primarily made up of composite particles in the size-grades studied.

ad 1) It has been shown (RIEZEBOS 1978) that the deposition of pyroxene-bearing ashes in the Valle de Ubaté-Chinquinquira where the Laguna de Fuquene is found, and very probably also in the Sabana de Bogotá, commenced at about 20000 BP. Between ca. 20000 BP and ca. 45000 BP only mica-, instead of pyroxene-bearing, ashes accumulated in this area. The development of the buried A horizons therefore very probably took place after ca. 20000 BP.

ad 2) Shards are found to be absent in the material of the buried A horizons which suggest that, in view of the fact that the supplied ashes originally were rich in volcanic glass particles (RIEZEBOS 1978), they have been removed physically (winnowing out) or chemically (strong weathering), before the buried A horizon development. It is generally known that the conditions in the surface zone constitute a much more effective environment for the destruction of shards, than the environmental conditions prevailing in a well-preserved ash layer included in a sedimentary column. It might be doubted, however, whether the surface environment during the Fuquene Stadial with its extremely dry and cold climate (VAN GEEL & VAN DER HAMMEN 1973) should have occasioned such an exhaustive destruction. At the most, these conditions might have contributed, before the final stabilization of the material by the buried A horizon, to a continuous winnowing out of the shards.

ad 3) Contradictory with a deduced physical removal of the shards seems to be the large abundances of composite particles in the material of the buried A horizons, and also in that of the C horizon of the surface soils. It has been shown that these "pseudo sand" particles owe their genesis and existence to cementation by allophanic material (RIEZEBOS & LUSTENHOUWER 1983). Generally it is believed that allophane originates in volcanic rocks mainly by dissolution of glass and felspars, and subsequent precipitation. Its genesis, therefore, requires probably more humid and warmer conditions than those inferred for the Fuquene Stadial (VAN GEEL & VAN DER HAMMEN 1973).

Glass components of andesitic tephra have been reported to alter into allophane in less than 300 years, and this allophane may persist for more than 40000 years without a subsequent change into crystalline mineral phases (KIRKMAN 1980). The origin of the allophanic cement of the abundant composite particles found in the buried and surface soils, may therefore date from a wetter period preceding the Fuquene Stadial as well as from the Guantiva interstadial (Fig. 3). However, the virtual lack of micas, constituting a substantial component of the ashes supplied before the Fuquene Stadial (RIEZEBOS 1978), does not speak in favour of the first possibility; unless the mica particles, and possibly also the shards, which escaped dissolution, have subsequently been removed from the ash materials during their exposure to reworking and redistribution in the Fuquene Stadial. Therefore, it might tentatively be concluded that the enrichment of the heavy minerals probably occurred during the El Abra Stadial and/or at the end of the Fuquene Stadial.

Concerning the second question holding whether these concentrations occurred contemporaneously in both soils or not, the presently available knowledge does not permit to make further speculations on that aspect.

5.2. CONCLUDING REMARKS

In stratigraphic and paleoenvironmental work surface and subsurface soils are increasingly utilized. Preserved buried soils may indicate by their actual presence, periods of non-deposition or phases of relative landscape stability. By means of their profile and material properties, soils may provide information on their environment of genesis and development. Their values as efficient stratigraphic and paleoenvironmental tools, however, is highly dependent upon the frequency, the nature, the intensity and the duration of the environmental changes during the development of the soils. As documents, soils become more mixed and the individual imprints of the environmental conditions more superimposed and/or obliterated, as the number of changes in climate, in surface configurations, and in the composition of supplied material becomes larger.

In areas of active explosive volcanism, often associated with tectonic activities, the induced geomorphological instabilities may be so large, that here in particular destruction and modification of evidence may widely occur. It is realised that man probably never will be able to recover completely the destroyed evidence. However, the use and application of methods, techniques and views uncommon in soil science, may possibly contribute to a better unravelment of the effects of the environmental influences, and in this way reduce the uncertainties of the interpretations.

ACKNOWLEDGEMENTS

This study was financially supported by the Netherlands Founation for the Advancement of Tropical Research (WOTRO). The manuscript has highly benefited from a careful review by Dr. D.H. Yaalon. My appreciation is also expressed to Mrs. O.M. Bergmeijer-de Vré, Mrs. M.C.G. Keijzer-v.d.Lubbe and Mr. J.v.d. Bliek, who assisted with the preparation of the manuscript.

REFERENCES

ANONYMOUS (1969): Geological Map of Colombia, Sheet K-11, Zipaquira, and Sheet K-10, Villeta. Instituto Nacional de Investigaciones Geologico-Mineras.
BOTERO, P.J. (1972): Soils of Guasca-Guatavita (Colombia). Unpublished MSc thesis, Institute of Aerial Survey and Earth Sciences, Enschede.
CORREAL-URREGO, G. & VAN DER HAMMEN, T. (1977): Investigaciones Arqueologicas en los Abrigos rocosos del Tequendama. Bibliotheca Banco Popular.
FRIED, G. (1983): Äolische Komponenten in Rotlehmen des Adamaua-Hochlandes / Kamerun. CATENA **10**, 87-97.

GRIGG, M.S. & RATHBUN, R.E. (1969): Hydraulic equivalence of minerals with consideration of the re-entrainment process. Prof. Pap. U.S. Geol. Surv. 650B, B77-B80.

HASEMANN, J.F. & MARSHALL, C.E. (1945): The use of heavy minerals in studies of the origin and development of soils. Research Bull. 387, Univ. of Missouri, 75 p.

JUNGERIUS, P.D. (1975): The properties of volcanic ash soils in dry parts of the Colombian Andes and their relation to soil erodibility. CATENA 2, 69-80.

JUNGERIUS, P.D. (1976): Quaternary landscape development of the Rio Magdalena Basin between Neiva and Bogotá, Colombia. A reconstruction based on evidence derived from paleosols and slope deposits. Palaeogr. Palaeoclimatol., Palaeoecol., 19, 89-137.

JUVIGNE, E. & SHIPLEY, S. (1983): Distribution of the heavy minerals in the downwind tephra lobe of the May 18, 1980 eruption of the Mount St Helens (Washington, U.S.A.). Eiszeitalter u. Gegenwart 33, 1-7.

KIRKMAN, J.H. (1980): Clay mineralogy of a sequence of andesitic tephra beds of Western Taranaki, New Zealand. Clay Minerals 15, 157-163.

LARSSON, W. (1937): Vulkanische Asche vom Ausbruch des Chilenischen Vulkans Quizapa (1932) in Argentinien gesammelt. Geol. Inst. Uppsala Bull., 26, 27-32.

RIEZEBOS, P.A. (1978): Petrographic aspects of a sequence of Quaternary volcanic ashes from the Laguna de Fuquene area, Colombia, and their stratigraphic significance. Quaternary Res., 10, 401-424.

RIEZEBOS, P.A. & LUSTENHOUWER, W.J. (1983): Characteristics and significance of composite particles derived from a Colombian Andosol profile. Geoderma, 30, 195-217.

SALLENGER, A.H. (1979): Inverse grading and hydraulic equivalence in grain-flow deposits. Journal of Sedim. Petrol., 49, 553-562.

SIDERENKO, A.V. (1956): Eolian differentiation of substances in the desert. Izv. Akad. Nauk SSSR, Ser. Geogr. 3.

SLINGERLAND, R.L. (1977): The effects of entrainment on the hydraulic equivalence relationships of light and heavy minerals in sands. Journal of Sedim. Petrol., 47, 753-770.

STEIDTMANN, J.R. (1982): Size-density sorting of sand-sized spheres during deposition from bedload transport and implications concerning hydraulic equivalence. Sedimentology, 29, 877-883.

VAN BEMMELEN, R.W. (1949): General geology of Indonesia and adjacent archipelagos. Govt. Printing Office, The Hague.

VAN GEEL, B. & VAN DER HAMMEN, T. (1973): Upper Quaternary vegetational and climatic sequence of the Fuquene area (Eastern Cordillera, Colombia). Palaeogeogr., Palaeoclimatol., Palaeoecol., 14, 9-92.

VAN DER HAMMEN, T. & GONZALEZ, F. (1963): Historia de clima y vegetación del Pleistoceno Superior y del Holoceno de la Sabana de Bogotá. Bol. Geol. (Bogotá), 11, 189-266.

VAN DER HAMMEN, T., WERNER, J.H. & VAN DOMMELEN, H. (1973): Palynological record of the upheaval of the Northern Andes: a study of the Pliocene and lower Quaternary of the Colombian Eastern Cordillera and the early evolution of its high-andean biota. Res. Palaeobot. Palynol., 16, 1-22.

VAN DER HAMMEN, T. (1978): Stratigraphy and environments of the Upper Quaternary of the El Abra corridor and rock shelters (Colombia). Palaeogeogr., Palaeoclimatol., Palaeoecol., 25, 111-162.

ZIMMERLE, W. (1973): Fossil heavy mineral concentrations. Geol. Rundschau, 62, 536-548.

Address of author:
P.A. Riezebos, Laboratory of Physical Geography and Soil Science, University of Amsterdam, Dapperstraat 115, 1093 BS Amsterdam, The Netherlands

E. Fernandez Caldas & Dan H. Yaalon (Eds):
VOLCANIC SOILS
CATENA SUPPLEMENT 7, Braunschweig 1985

THE WEATHERING OF BASALT
IN AN ARCTIC ENVIRONMENT

L.J. **Evans** and W. **Chesworth,** Ontario

SUMMARY

The weathering of basalt was investigated by studying five soil profiles developed on Tertiary basalts at Cape Dyer, Baffin Island, Canada. The mean annual soil temperature of the area is $-10.3°C$ and the precipitation 663 mm, with over 6 m of snow. Chemical and mineralogical analyses were performed on rock, soil and soil separates. Results indicated that weathering of olivine and augite had occurred but was not extensive, and that there was a relative increase in the content of plagioclase feldspar as the particle size decreased to fine silt. The clay fraction contained between 22-27% allophane, with dioctahedral vermiculite as the major phyllosilicate mineral present. Contents of Na_2O and K_2O in the silt, and the presence of quartz in the fine sand and silt fractions, suggested that some eolian additions of granitic material had occurred from the surrounding Precambrian felsic rocks.

1. INTRODUCTION

The weathering of basalt has been studied in a variety of climates, particularly in subtropical and tropical areas. Results of these studies indicate that smectite and kaolinite are the major phyllosilicate minerals present in the clay fractions, with halloysite occurring in lesser amounts. Kaolinite predominates in well drained sites, in acidic soils and in areas of high rainfall (SHERMAN & UEHARA 1956, SINGER 1966, SWINDALE 1966). The composition of the smectite is generally saponitic or nontronitic (WEAVER & POLLARD 1973), or more rarely intermediate between di- and trioctahedral (CURTIN & SMILLIE 1981). Less common phyllosilicate minerals reported are vermiculite (SMITH 1957, BRINER & JACKSON 1970) and interstratified montmorillonite-vermiculite-illite (BAIN & RUSSELL 1980). Amorphous aluminosilicates, such as allophanes, are frequent constituents of clay fractions from weathered basalts (SIEFFERMAN & MILLOT 1969).

Studies on the weathering of basalt in environments with pergelic soil temperature regimes (MAST $< 0°C$) are rare. CAMPBELL & CLARIDGE (1968) reported that basaltic soils in the vicinity of Edisto Inlet, Antarctica, contained clay fractions that were essentially amorphous, although no allophane was detected. In some tundra soils of the USSR, SOLOKOV & GRADUSOV (1979) reported trioctahedral smectite and chlorite as the major weathered products of basalt in soils with a moss-lichen-shrub cover. We report here the results of a study on the chemical and mineralogical composition of weathered basalt on Baffin Island, N.W.T., Canada.

ISSN 0722–0723 / ISBN 3–923381–06–9

2. DESCRIPTION OF AREA AND FIELD SAMPLING

2.1. BEDROCK GEOLOGY AND CLIMATE

Basalts of early Tertiary age outcrop at Cape Dyer, Baffin Island, N.W.T., Canada. These rocks have been correlated with similar basaltic outcrops on the west coast of Greenland (CLARKE 1970). The basalts consist primarily of picritic and olivine-rich flows, with average thicknesses of 3.5 m (CLARKE & UPTON 1971). The upper portion of the flows, in which the soils have developed, consists of a red coloured amygdaloidal basalt, with the vesicles commonly filled with zeolites. The lack of glacial striae and erratics on the exposures suggests an absence of glacial erosion (IVES 1963).

The nearest climatological station to the sampling areas is at Cape Dyer (66°35'N, 61°37'W), approximately 5 km distant. Recorded mean annual temperatures average $-10.3°C$, with mean summer temperatures (June, July and August) of $3.3°C$. Extreme highs may reach $19°C$ and extreme lows $-47°C$. Precipitation averages 663 mm, with over 6 m of snow recorded annually.

Tab. 1: GENERAL CHARACTERISTICS OF THE STUDIED SOILS

| | | | | | | <2 mm fraction | | | |
	Depth (cm) (cm)	pH	C %	Gravel %	Coarse Sand %	Fine Sand %	Coarse Silt %	Fine Silt %	Clay %
Profile 1 (10YR 4/3; weak, coarse sub-angular blocky)									
ACy1	0-20	5.6	0.33	35	34.2	30.6	22.6	11.5	1.2
ACy2	21-40	5.7	0.31	50	38.4	33.0	19.1	8.3	1.1
Profile 2 (5YR 3/4; weak, medium to coarse sub-angular blocky)									
ACy1	0-20	6.3	0.27	40	38.4	22.1	21.3	15.1	3.0
ACy2	20-40	6.2	0.30	49	32.5	24.5	25.9	14.2	2.8
Profile 3 (7.5YR 4/6; structureless)									
ACy1	0-20	6.3	0.34	59	41.2	24.5	16.9	14.0	3.5
ACy2	20-40	6.3	0.25	57	41.2	25.2	17.6	13.3	2.7
Profile 4 (10YR 4/4; structureless to weak coarse sub-angular blocky)									
ACy1	0-20	6.3	0.23	61	42.9	19.2	19.1	16.2	2.6
ACy2	20-40	6.1	0.20	67	78.9	6.9	5.3	7.3	1.6
Profile 5 (10YR 5/3; structureless)									
ACy	0-20	6.3	0.26	32	26.1	34.2	15.4	21.5	2.7

2.2. FIELD SAMPLING AND SOIL DESCRIPTIONS

The bedrock exposures vary in elevation from 500 m to 850 m and consist largely of boulders ranging in diameter from 10 to 50 cm. Vegetative cover is virtually non existant, except for occasional plants of arctic heather (Cassiope hypnoides). Areas containing soils constituted less than 5% of the exposures and were generally found in lower lying depressions.

These areas showed evidence of intense cryoturbation and contained many sorted nets, polygons and circles, varying in diameter from less than 1 m to 3 m. Permafrost was generally encountered at about 40 cm from the soil surface and bedrock at about 50 cm.

In total, nine soil horizons were sampled from five soil profiles. The soil ACy horizons were divided into two subhorizons (0-20 and 20-40 cm) if permafrost was deeper than 40 cm. The soils contained considerable contents of gravel and had sandy loam to loamy sand textures, clay contents less than 4%, organic C contents between 0.20 and 0.34% and pH's between 5.6 and 6.3 (Table 1).

The soils were classified as Regosolic Turbic Cryosols (CANADA SOIL SURVEY COMMITTEE 1978) or Pergelic Cryorthents (SOIL SURVEY STAFF 1975).

3. ANALYTICAL PROCEDURE

Air-dry samples were sieved through 2000 μm sieves and the fine earth fraction (< 2000 μm) and gravel separated. Coarse sand (2000-200 μm) and fine sand (200-50 μm) were separated from the fine earth by sieving, and the coarse silt (50-200 μm), fine silt (20-2 μm) and clay (< 2 μm) by sedimentation.

pH's were measured in 2:1 soil:0.01M $CaCl_2$ suspensions and organic C contents using a Leco induction furnance. Al, Fe, Mg, Ca, K, Na, and Mn were determined by atomic absorption spectroscopy and Ti by an auto-analyser technique using tiron after digestion of the samples in HF; Si was determined using an auto-analyser technique utilizing the molybdate blue reaction after fusion of the samples in Na_2CO_3. FeO contents were determined by the colorimetric method of WILSON (1960). H_2O contents were determined as the weight loss under N_2 between 110-950°C using a Stanton Redcroft Thermobalance. Dithionite-citrate-bicarbonate (DCB) extractable Fe, Al and Mn, and NaOH extractable Si and Al were determined on the clay fraction after the methods of ALEXIADES & JACKSON (1965). The cation exchange capacities of the clay fractions were determined by saturating the exchange complex with Cu^{2+} and determining the total content of Cu by atomic absorption spectroscopy after HF digestion.

The mineralogical compositions of the rock, soil and soil separates were examined by X-ray diffraction using Ni-filtered Cu radiation on a Rigaku D-Max X-Ray Diffractometer. Rocks, soils, sands and silts were examined as randomly-orientated pressed powders and the treated clays as orientated films on glass slides after saturation with Mg and K. The Mg slides were also examined after ethylene glycol and glycerol solvation and the K-slides after heating to 500°C. Thin sections of randomly selected rock samples from each of the soil profiles were examined under the petrological microscope.

4. RESULTS

4.1. CHEMISTRY AND MINERALOGY OF ROCK AND SOIL FINE EARTH

Rock samples

The mineralogy of the soil parent material was investigated by both petrological techniques and by powder X-ray diffraction analyses of rock samples. Eighteen rock samples were collected as stones, approximately 5-10 cm across, taken from each of the soil horizons - two per horizon.

Examination of rock thin sections revealed that the rock ground mass consisted largely of plagioclase feldspar, olivine and augite, with between 5-10% of the olivine existing as phenocrysts, often altered to chlorite and/or iddingsite on their rims. Occasional laths of plagioclase and opaque magnetite and/or ilmenite were also observed. These observations are very similar to those reported by MACLEAN et al. (1978) for basaltic cores taken from Baffin Bay, approximately 33-89 km south of Cape Dyer.

Examination of powder X-ray diffraction patterns confirmed the presence of calcium-rich plagioclase, olivines and augite. Amygdaloidal zeolite present in some samples was identified by X-ray diffraction analysis as chabazite and analyses of the d(130) X-ray diffraction peaks of olivine in coarse and fine sand fractions indicated that the composition of the olivine was Fo_{76}-Fo_{84}, using the method of YODER & SAHAMA (1957).

The chemical composition of the eighteen rock fragments analysed (table 2) was consistent with a mineralogical composition consisting of approximately 40% feldspar (principally anorthite-rich plagioclase) and 55% olivines and pyroxenes (principally forsteritic-olivine and augite).

Tab. 2:   AVERAGE CHEMICAL COMPOSITION OF ROCK AND SOIL SAMPLES

	SiO_2	Al_2C_3	FeO	Fe_2O_3	CaO	MgO	K_2O	Na_2O	MnO	TiO_2	H_2O
Rock	47.55	15.22	5.43	6.68	8.80	11.26	0.08	1.39	0.19	0.98	1.83
Soil	48.39	15.67	5.75	5.50	7.49	10.52	0.22	1.98	0.17	0.97	2.30

Soil fine earth fractions

Examination of X-ray diffraction traces of powered soil fine earth fractions indicated lower contents of olivine and augite, and higher contents of plagioclase feldspars than the rock samples. The composition of the soil samples was also different in containing small amounts of quartz.

Relative to rock samples, the soil samples had higher contents of SiO_2, Al_2O_3, K_2O and Na_2O, but lower contents of CaO, MgO, MnO and total Fe, again suggesting that weathering of olivine and augite has occurred.

4.2.   CHEMISTRY AND MINERALOGY OF SOIL SEPARATES

Sand fractions

X-ray diffraction analyses indicated the presence of plagioclase feldspar, olivine and augite in both coarse and fine sand fractions, with feldspar predominating in all fractions. The mineralogy of the fractions was very similar and this was confirmed by the similarity in their chemical compositions (table 3). However there were some differences in that the fine sand fractions of horizons 1Ahy2 and 2Ahy2 contained small amounts of chlorite and the fine sand fractions of horizons 1Ahy1, 1Ahy2, 2Ahy1, 2Ahy2, 4Ahy1 and 4Ahy2 contained quartz. Quartz and chlorite did not however occur in any of the coarse sand fractions.

Tab. 3: AVERAGE PARTIAL CHEMICAL COMPOSITION OF SAND AND SILT FRACTIONS

	Al_2O_3	FeO	Fe_2O_3	CaO	MgO	K_2O	Na_2O	MnO	TiO_2
Coarse sand	13.09	6.30	6.00	7.55	12.56	0.13	1.21	0.18	0.88
Fine sand	12.22	6.74	5.36	7.12	12.70	0.23	1.23	0.18	1.00
Coarse silt	18.06	5.46	4.95	8.86	8.84	0.21	1.51	0.17	1.06
Fine silt	24.58	2.72	5.06	5.97	4.60	0.34	1.54	0.11	1.08

Silt fractions

The silt fractions contained considerably less augite and olivine than the sand fractions. Whereas augite could be identified in all coarse silt fractions, olivine was only identified in four. Similarly augite was identified in only two fine sand fractions, whereas augite was identified in six. Quartz was identified in both coarse and fine silt fractions of horizons 1Ahy1, 1Ahy2, 2Ahy2, and in the fine silt fraction of horizon 3Ahy1. Chlorite occurred in all fine silt fractions and in four coarse silt fractions.

Chemically the coarse and fine silt fractions differed from each other in that the content of CaO, MgO, MnO and total Fe decreased with a decrease in particle size, whereas the content of Al_2O_3, K_2O, Na_2O and TiO_2 increased (table 3). These chemical changes presumably reflect the decrease in contents of olivine and augite and an increase in the content of plagioclase feldspar.

Clay fractions

Considerable amounts of material were dissolved from the clay fractions by successive extractions with dithionite-citrate-bicarbonate (DCB) and NaOH (table 4). Contents of Fe_2O_3 dissolved by DCB ranged from 2.7 to 4.9%, Al_2O_3 from 1.6 to 4.7% and MnO from 0.19 to 0.27%. NaOH-extractable SiO_2 and Al_2O_3 ranged from 10.6 to 15.0% and 8.6 to 11.0% respectively. Contents of amorphous aluminosilicates ranged from 21.4 to 27.7% with a mean molar Si/Al ratio of 2.3. These contents were calculated using the formula given by ALEXIADES & JACKSON (1965).

The major phyllosilicate mineral found in all the clay fractions was vermiculite, with only minor, trace or negligible amounts of illite and chlorite. Identification of phyllosilicates was made after removal of iron oxide/hydroxide coatings and amorphous aluminosilicates. Vermiculite was identified by the collapse of the 1.4 nm peak of Mg-saturated clays to 1.0 nm on K-saturation and the non-expansion of this peak on solvation with glycerol. For all the clay samples, there was some expansion to a greater or lesser extent to 1.6 nm on solvation

Tab. 4: AVERAGE EXTRACTABLE Si, Al, Fe and Mn FROM CLAY FRACTIONS

Dithionite-citrate-bicarbonate			Boiling 0.2 M NaOH		
Al_2O_3 %	Fe_2O_3 %	MnO %	SiO_2 %	Al_2O_3 %	Si/Al ratio
2.88	3.61	0.21	13.54	9.99	2.30

with ethylene glycol. Plagioclase feldspars were detected in all the clay fractions, whereas olivine and augite were not.

To calculate a structural formula for the vermiculite in the clay fractions, the contents of CaO and Na_2O in the clay were allocated to anorthite and albite respectively, the content of K_2O allocated to illite (assuming an ideal formula of muscovite) and the content of TiO_2 allocated to rutile. The calculated amounts of feldspars present in the clay fractions were 14% plagioclase feldspar ($Ab_{43}An_{57}$) and the calculated amount of illite was 5%. The requisite amounts of SiO_2, Al_2O_3, CaO, K_2O, Na_2O and TiO_2 ascribe to the feldspars, illite and rutile were then subtracted from the average chemical composition of the treated clays given in table 5.

The chemical composition of the clay was then adjusted for the loss of these constituents. The structural formula of the vermiculite was calculated assuming a cationic valency of 44. This formula ignores the presence of trace amounts of chlorite in some samples. Results from these calculations showed that the vermiculite present in the clay fraction was dioctahedral with a layer charge of 0.83. The formula of the vermiculite was calculated as:

$$(Si^{4+} \quad Al^{3+}\ )(Al^{3+} \quad Fe^{3+} \quad Fe^{2+} \quad Mg^{2+} \quad Mn^{2+}\ )0 \quad (OH) \quad Cu$$
$$\ 7.273 \quad 0.727 \quad 2.889 \quad 0.005 \quad 0.731 \quad 0.86 \quad 0.007 \ 20 \quad 4 \ 0.412$$

The layer charge of 0.829 is neutralized by 0.824 charges contributed by the Cu. This charge corresponds to a cation exchange capacity (cec) of 110 me/100 g. The average cec of the treated clay fractions determined by exchanging the Cu^{2+} ions with NH_4^+ amounted to 76.8 meq/100 g, or 96.1 meq/100 g after correcting for the content of feldspars, illite and rutile. Thus the calculated cec although a little higher than the measured cec suggests that the calculated formula is probably realistic in general terms given the assumptions involved in its calculation.

The absence of chlorite in the clay fractions suggests that the chlorite present in silt and sand fractions has resulting from post-depositional alteration of the basalt and not from processes associated with pedogenesis.

Tab. 5:   CHEMICAL COMPOSITION OF TREATED AND UNTREATED CLAYS

	SiO_2	Al_2O_3	FeO	Fe_2O_3	CaO	MgO	K_2O	Na_2O	MnO	TiO_2	CuO*	H_2O
Clay (untreated)	47.37	26.14	1.80	3.65	1.27	2.15	0.43	0.43	0.26	0.52	3.53	10.61
Clay (treated)	50.78	23.28	4.89	0.13	1.68	3.22	0.55	0.75	0.05	0.80	3.06	9.67

*   Clay samples were saturated with Cu^{2+} before analysis

4.   DISCUSSION

The general trend in a weathering system is towards compositions enriched in SiO_2, Al_2O_3 and Fe_2O_3 relative to all other components (CHESWORTH 1973a, 1973b). Our data plotted in the SiO_2-Al_2O_3-Fe_2O_3 triangle (figure 1a) indicates little or no change in chemistry from rock to soil. This of course is not too surprising given the climatic regime in which the

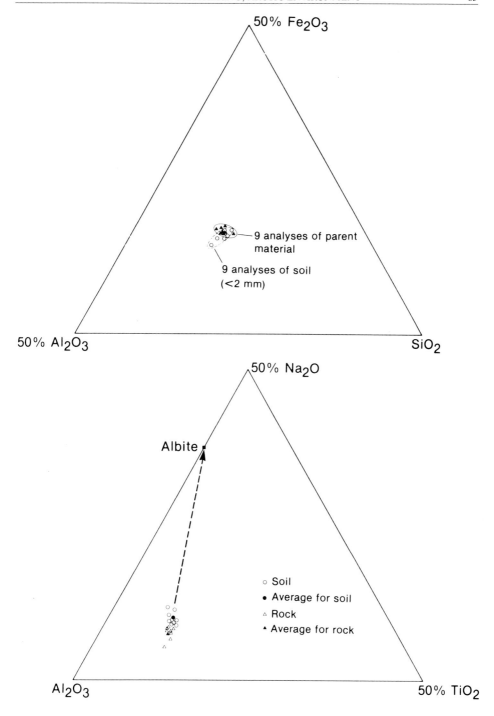

Fig. 1: Average chemical composition of soil and rock samples.
a) plotted in Al_2O_3 - Fe_2O_3 - SiO_2 co-ordinates; b) plotted in Al_2O_3 - Na_2O - TiO_2 co-ordinates

weathering processes are taking place.

However, examination of the original data (table 2) reveals an increase in alkalis, particularly Na, from rock to soil. This would not be expected as a consequence of weathering. Two mechanisms for enrichment of Na that immediately suggest themselves are – a) contamination of the basaltic soils by eolian addition from the surrounding granitic material and b) addition of marine salts.

Precambrian granites outcrop at elevations of 700-800 m along the Arctic coastline of Cape Dyer and essentially surround the basaltic outcrops. Boulders of the granite are pitted by tafoni, good evidence for the active role of wind in the region. Thus a contaminant exists and a mechanism to transport it. The second possibility, that of addition of marine salts as aerosols, has been recognised elsewhere (YAALON & LOMAS 1970, POTTS & EVANS 1977).

An aid in selecting between these two mechanisms is to represent the data in a graphical manner. This technique involves choosing two components likely to be added to the system – 1) by the contaminating source and – 2) in different proportions by the two possible sources. In this case Na_2O and Al_2O_3 serve this purpose. From the granite they will be added in a ratio likely to be present in alkali feldspars. From a marine aerosol only Na_2O will be added. If now a third component is taken, one that is unlikely to change significantly during the addition of either contaminant, all three components may be represented in a triangular plot (figure 1b) to investigate whether the contamination trend evolves towards an alkali feldspar composition or the Na_2O apex of the diagram. As the third component on our triangular representation we have chosen TiO_2.

Clearly the trend (figure 1b) of the average soil and rock samples is close to being in line with the albite composition, a clear indicator of contamination by granitic material. The projection does appear however to be slightly on the high Na_2O side of albite, indicating that aerosol addition is still possible, though of secondary importance.

The graphical analysis, and the conclusion that granitoid material has been the main source of contamination in the system, helps to explain some of the mineralogical observations. The fine sand and silt fractions of some of the samples contain a small amount of quartz, a mineral not present in the original basalt. In addition, dioctahedral vermiculite is present in the clay fractions. Dioctahedral vermiculite is a well recognised phase produced in the early stages of the weathering of acid rocks and vermiculite has been identified in soils developed from granitic material near Broughton Island, about 150 km northwest of the study area (EVANS & CAMERON 1979).

Not surprisingly, the only cations that show a loss in soil relative to the parent material, i.e. display a weathering trend, are Mg and Ca. These elements are present predominantly in olivine, augite and calcic plagioclase – the dominant minerals in the basalt. Some Mg and Ca could be added by contaminating granitic material, but, unlike the alkalis, the amount present in the contaminants would be much less than in the basalt, and would not be enough to mask the predominant weathering trend. However the relative changes in such components are not large, again confirming that we are dealing with a very immature weathering system.

REFERENCES

ALEXIADES, C.A. & JACKSON, M.L. (1967): Quantitative analysis of soils and sediments. Clays Clay Minerals, **14**, 35.

BAIN, D.C. & RUSSELL, J.D. (1980): Swelling minerals in a basalt and its weathered products from Morvern, Scotland. I. Interstratified montmorillonite-vermiculite-illite. Clay Minerals, 15, 445.

BRINER, G.P. & JACKSON, M.L. (1970): Mineralogical analysis of clays in soils developed from basalt in Australia. Israel J. Chemistry, 8, 487.

CANADA SOIL SURVEY COMMITTEE (1978): The Canadian system of soil classification. Can. Dept. Agric. Publ. 1646, Ottawa, Ont., 165.

CAMPBELL, I.B. & CLARIDGE, G.G.C. (1968): Soils in the vicinity of Edisto Inlet, Victoria Land, Antarctica. New Zealand J. Science, 11, 498.

CHESWORTH, W. (1973a): The parent rock effect in the genesis of soil. Geoderma, 10, 215.

CHESWORTH, W. (1973b): The residua system of chemical weathering: a model for the chemical breakdown of silicate rocks at the surface of the earth. J. Soil Science, 24, 69.

CLARKE, D.B. (1970): Tertiary basalts of Baffin Bay: possible primary magma from the mantle. Contr. Mineral. Petrol., 25, 203.

CLARKE, D.B. & UPTON, B.G.J. (1971): Tertiary basalts of Baffin Island: field relations and tectonic setting. Canadian J. Earth Science, 8, 248.

CURTIN, D. & SMILLIE, G.W. (1981): Composition and origin of smectite in soils derived from basalt in Northern Ireland. Clays Clay Minerals, 29, 277.

EVANS, L.J. & CAMERON, B.H. (1979): A chronosequence of soils developed from granitic morainal material. Baffin Island, N.W.T. Canadian J. Soil Science, 59, 203.

IVES, J.D. (1963): Field problems in determining the maximum extent of Pleistocene glaciation along the eastern Canadian seaboard – a geographers point of view. In: North Atlantic biota and their history. Pergamon press, London, 337.

MACLEAN, B., FALCONER, R.K.H. & CLARKE, D.B. (1978): Tertiary basalts of western Davis Strait: bedrock core samples and geophysical data. Canadian J. Earth Science, 15, 772.

POTTS, M.J. & EVANS, L.J. (1977): The possible effect of air-borne salt of marine origin on the weathering of illite in west Wales. Clay Minerals, 12, 359.

SHERMAN, C.D. & UEHARA, G. (1956): The weathering of olivine basalt in Hawaii and its pedological significance. Soil Science Society of America Proceedings, 20, 337.

SIEFFERMAN, G. & MILLOT, G. (1969): Equatorial and tropical weathering of recent basalts from Cameroon: allophanes, halloysite, meta-halloysite, kaolinite and gibbsite. International Clay Conference, Tokyo, 417.

SINGER, A. (1966): The mineralogy of the clay fraction from basaltic soils in the Galilee Israel. J. Soil Science, 17, 136.

SMITH, J. (1957): A mineralogical study of weathering and soil formation from olivine basalt. J. Soil Science, 8, 225.

SOIL SURVEY STAFF (1975): Soil Taxonomy. Agric. Handb. no. 436, SCS–USDA. U.S. Government Printing Office, Washington, D.C., 754 p.

SOLOKOV, I.A. & GRADUSOV, B.P. (1979): Soil formation and weathering on basic rocks in a cold humid climate. Soviet Soil Science, 10, 11.

SWINDALE, L.D. (1966): A mineralogical study of soils derived from basic and ultrabasic rocks in New Zealand. New Zealand J. Science, 9, 484.

WEAVER, C.E. & POLLARD, L.D. (1973): The chemistry of clay minerals. Elsevier, Amsterdam.

WILSON, A.D. (1960): The micro-determination of ferrous iron in silicate minerals by a volumetric and a colorimetric method. Analyst, 85, 823.

YAALON, D.H. & LOMAS, J. (1970): Factors controlling the supply and the chemical composition of aerosols in a near-shore and coastal environment. Agricultural Meteorology, 7, 443.

YODER, H.S. & SAHAMA, T.G. (1957): Olivine X-ray determinative curve. American Mineralogist, 42, 475.

Address of authors:

L.J. Evans and Ward Chesworth, Department of Land Resource Science, University of Guelph Ontario, N1G 2W1, Canada

E. Fernandez Caldas & Dan. H. Yaalon (Eds):
VOLCANIC SOILS
CATENA SUPPLEMENT 7, Braunschweig 1985

CARBONATISATION AS A SOIL FORMING PROCESS ON SOILS FROM BASIC PYROCLASTIC FALL DEPOSITS ON THE ISLAND OF LANZAROTE, SPAIN

R. **Jahn**, Th. **Gudmundsson** and K. **Stahr**, Berlin

SUMMARY

On the island of Lanzarote carbonatisation and soil development were studied in three landscapes of different ages; landscape IVB is about 250 years old, IVA subrecent and III early young pleistocene. On all the surfaces the soils have developed on basic to ultrabasic pyroclastic fall deposits or basalt. The carbonatisation on IVB is in initial stadium with ca. 1 kg $CaCO_3 m^2$ accumulated whereas IVA with > 10 kg/m^2, III with > 100 kg/m^2 and II with > 1000 kg/m^2 have destinctive carbonate accumulation horizons.

In the older pyroclastic fall deposits the carbonate accumulates in numerous horizons to depths of at least 4 m, and are frequently associated with plant roots. These carbonate accumulation horizons also show a higher degree of weathering, indicated by finer texture and higher solubility of Fe, Al, Si and Mn, than the surrounding horizons. Attempts to calculate losses and gains of Ca show that only about 1/3 of the Ca released through weathering is bound as calcium carbonate in the soil. Thus, the main source of Ca will be the parent material.

ZUSAMMENFASSUNG

In drei unterschiedlich alten Landschaften Lanzarotes wird die Carbonatisierung und Bodenentwicklung studiert. Das Alter dieser Landoberflächen ist durch die jeweiligen basaltischen Eruptiva IVB 250 Jahre alt, IVA subrecent, III jüngeres Jungpleistozän, definiert. Die Carbonatisierung auf Fläche IVB ist in einem Initialstadium mit ca. 1 kg $CaCO_3$ – Akkumulation/m^2, während die Kalkanreicherung in IVA > 10 kg/m^2 und in III > 100 kg/m^2 erreicht.

Auf den noch älteren Landschaften der Serien II und I kann die Carbonatakkumulation > 1000 kg/m^2 erreichen.

In den pyroklastischen Materialien der Serien IVA und III akkumuliert das Carbonat in zahlreichen Horizonten bis zu einer Tiefe von mindestens 4 m und sind teilweise durchwurzelt. Diese Horizonte zeigen gegenüber ihrer Umgebung eine höhere Verwitterungsintensität, welche durch feinere Textur und höhere Gehalte an löslichen Fe, Al, Si und Mn gekennzeichnet ist. Der Versuch, die Ca-Verluste und -Anreicherungen zu kalkulieren, zeigt, daß nur ca. 1/3 des durch Verwitterung freigesetzten Calciums im Profil als $CaCO_3$ gebunden wird. Somit kann als Hauptquelle der Carbonatanreicherung das Ausgangsmaterial bestimmt werden.

1. INTRODUCTION

The island Lanzarote is formed from volcanic eruptive rocks, which build up distinct landsurfaces of different ages. This offers a unique opportunity to study a chronosequence of soils. As the volcanic rocks originate from distinct eruptive phases, time as a soil forming factor is determined by these phases.

ISSN 0722–0723 / ISBN 3–923381–06–9

The other factors show only minor variations, although alternations of the climate during the course of time do in fact influence the strict time dependence.

The age of the six basaltic series covers the period from miocene until the catastrophic eruptions of 1730-36 and 1824 (Instituto Geologico 1967-1968). The geochemistry of the rocktypes is basic to ultrabasic with small variations (FUSTER et al. 1968).

The climate of the island (800 km², 29°N) is characterized by an average annual precipitation of ~ 140 mm (< 100 - 250 mm) and mean annual temperatures of ~ 20°C. The rainfall has its maximum between November and February. The climate is classified as semiarid with warm temperature (BWh to BSk after KÖPPEN). The prevailing soil moisture regimes are ustic tending to aridic at low elevations and xeric for the higher elevations. The corresponding soil temperature regimes are hyperthermic and higher up thermic.

The soils of the older series are characterized by severe erosion and deposition of soil material in depressions. The more recent soils of the younger series III, IVA and IVB are less or not affected by erosion (compare table 1). Therefore, these surfaces are particulary suitable for studying soil development in relation to time.

JAHN et al. (1983) showed that the progress of some soil forming processes is time dependent but that the intensity of the different processes varies strongly (compare Figure 1).

The purpose of this paper is to describe the extent and morphology of the carbonate accumulation in relation to other soil forming processes in the younger soils.

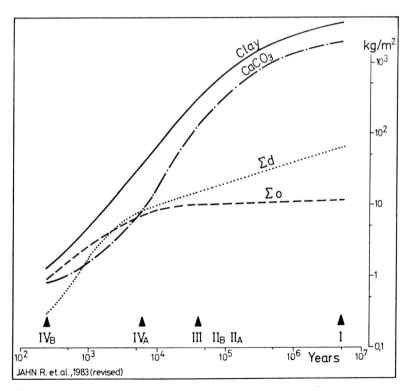

Fig. 1:   Time dependence of soil forming processes (from JAHN et al., 1983).

2. MATERIALS AND METHODS

2.1. SOILS

The age, parent material, soil type and location of the analysed profiles is presented in Table 1.

Tab. 1: THE ANALYSED SOIL PROFILES IN RELATION TO THE GEOLOGICAL FORMATIONS OF LANZAROTE

Geological formation Age	Rocktype	Soiltype FAO	USDA	Nr. and location	
IVB 250 years	Lapilli & Cinders	eutric Regosol	Xerorthent	IVB-101	Montaña Negra
	Ashy Lapilli &Cinders	eutric Regosol	Xerorthent	IVB-313	Caldera Blanca
IVA subrecent	Lapilli & Cinders	mollic Andosol	Haploxeroll	IVA-605	Los Helechos
	Lapilli & Cinders	mollic Andosol	Haploxeroll	IVA-660	Monte Corona
III younger early	Lapilli & Cinders	luvic Xerosol	Haploxeralf	III-510	El Peñón
pleistocene	Lapilli & Cinders	calcic Xerosol	Xerochrept	III-0200	Lomo de Camacho
	Basalt	luvic Xerosol	Haploxeralf	III-570	Eremita de los Dolores

2.2. METHODS

The analysed profiles were selected from the data collected after mapping two areas in each of the six different landscapes of Lanzarote.

The chemical analyses were carried out using the methods of SCHLICHTING & BLUME (1964) with the following modifications:
–   The extraction with ammonium-oxalate oxalic acid mixture at pH 3,25, room temperature and in darkness according to TAMM & SCHWERTMANN.
–   The extraction with sodium dithionite-citrate at pH 7,3, according to MEHRA & JACKSON.
–   Conductimetric analyses of carbonates after digestion with H_3PO_4 at 80°C using WÖSTHOFF-apparatus.
–   HF–HClO$_4$ dissolution according to JACKSON for total elemental analyses. The elements were determined using atom absorption equipment. Prior to the Ca determination 1% La-solution was added.
–   Particle size analysis according to SCHLICHTING & BLUME (1964) after H_2O_2 and HCl treatment. The clay content is computed taking the $CaCO_3$ and OM contents into account.

Data of Ca and $CaCO_3$ are calculated to volume of soil, using analysis of total Ca, carbonate Ca and CO_2 in fine earth and stones, fine earth-/stone content and bulk density.

Undisturbed soil samples were impregnated in polyester resin, and the prepared thin sections described according to FITZPATRICK (1980) using the calcite terminology of WIEDER & YAALON (1983).

Tab. 2: MACRO– AND MICROMORPHOLOGY OF THE CARBONATE ACCUMULATION
HORIZONS AND THE AMOUNT OF CARBONATES ACCUMULATED IN THE THREE
YOUNGEST LANDSCAPES OF LANZAROTE

IVB pyroclastic fall deposit	IVA pyroclastic fall deposit	III pyroclastic fall deposit
	Carbonate accumulation	
$\sim$ 1 kg $CaCO_3m^2$	> 10 kg $CaCO_3/m^2$	> 100 kg $CaCO_3/m^2$
	Macromorphology	
About 10 cm below the surface an approximately 2 cm thick Ck-horizon occurs, the lapilli is slightly cemented and enriched with fine-earth.	Accumulation in numerous very thin (2-5 cm) Ck-horizons from 30 to 300 cm depth. The Ck-horizons frequently have two 0.5 to 1,5 mm thick hard shells facing each other in the center of the horizons. Predominantly, a root mat or traces of root mat is located between the shells. Vertical pipes of carbonate accumulation 5-15 cm in diameter are frequent from 30 to 150 cm depth.	The main accumulation is in a Ckm-horizon, 40-60 cm thick, below about 100 cm depth. The horizon has a coarse laminar to blocky structure. The big structural units frequently have a fine laminar shell on the surface and they usually also show a horizontal layering where carbonate-rich soil and carbonate-impregnated lapilli alternate. Below the main accumulation, there are numerous very thin horizons as in IVA. Vertical pipes 5-20 cm in diameter are prominent from the main Ckm-horizon down to 150-200 cm depth.
	Micromorphology	
The large to very large lapilli particles are partly enclosed, and partly bridged by the light brown weakly anisotropic matrix. The calcite forms microcalcites in the matrix, but microsparites and sparites in enclosed pores of the lapilli.	In the hard shells in the center of the horizons the lapilli has been displaced by light brown to yellowish-brown matrix. The matrix is dominantly banded with carbonate-rich light brown anisotropic and weakly carbonized yellowish brown isotropic to weakly anisotropic alternating bands. The bands are horizontally oriented and are smooth to wavy. The carbonate-rich bands frequently consist of poorly developed segregations or concretions. Close to the shells the medium to very large lapilli is embedded in light brown, dark brown and yellowish-brown matrix. Here, the matrix is also banded but the bands are undulating and discontinuous due to the lapilli particles. In the matrix, there are occasional discrete spherical to ovoid, subrounded to well rounded, medium to small pores. These pores are usually linearly distributed in the banded matrix. The calcite consists of microcalcite in the matrix and of microsparites and sparites in enclosed pores of the lapilli. The lapilli is dominantly fresh to nearly fresh, the alteration being katamorphous, pellicular and dotted.	In the main Ckm-horizon, there are zones where the lapilli has predominantly been displaced by light brown calcic matrix or calcic concretions. The calcic matrix and concretions form a more or less continous phase with occasional discrete pores which are often connected by white or pale grey calcic matrix, indicating that they form channels in which the calcite has precipitated. In other zones, the lapilli has been cemented by light brown calcic matrix or calcic concretions with other basic features as described above. The thin horizons below the main accumulation are characterized by a center where the light brown calcic matrix or calcic concretions have displaced the lapilli. The matrix may be banded as in IVA or it forms a continuous phase. The vertical pipes have a concentric structure with a small to medium pore in the center surrounded by light brown weakly anisotropic, weakly to moderately calcic matrix. This grades gradually into the lapilli cemented by the same type of matrix. The calcite consists of microcalcite in the matrix and of microsparites and sparites in enclosed pores of the lapilli. The degree of alteration of the lapilli ranges from fresh to completely altered, the alteration being katamorphous, pellicular or dotted.

3. RESULTS AND DISCUSSION

3.1. DEGREE OF CARBONATISATION IN SOILS OF DIFFERENT AGE

A conspicious feature of the north-eastern Canary Islands and the drier parts of the others is that the older landsurfaces are dominantly covered with thick carbonate accumulations. The calcrete has been exposed in these vast areas due to erosion of the upper part of the soil.

The main sources of calcium in the carbonate accumulation are the basic to ultrabasic basalts and pyroclastic deposits. This basaltic parent material contains 5 to 7% calcium, which means that up to 500 kg $CaCO_3$ may be formed from the weathering of 1 m^3 basalt. Assuming that inputs of Ca through precipitation and sea spray are less than 2 $g/m^2/year$ ($\cong$ 4 l sea spray/m^2/year), the weathering of basalt will be the main source of Ca. Recent pyroclastic deposits may contain small amounts of calcrete fragments from older surfaces, which have been broken and carried by eruptions.

The differences in amount and nature of the enriched calcium carbonate depending on their age are best observed in the landscapes IVB to III, as here the profile differentiation takes place and most of the profiles are still autochthonous.

The two features that characterize the development of these soils are the amount of newly formed carbonates as well as the morphology of the enriched horizons. The most striking characteristics are summarized in Table 2. A conspicious feature in the profiles on landscapes IVA and III are very thin (2 - 4 cm thick) bands of carbonate accumulation (GUDMUNDSSON et al. 1984). These horizons contain more fine material and organic matter than the horizons above and below. Frequently, the center of these horizons have fine laminar shells or pans that are 0,5 to 1,5 mm thick. The $CaCO_3$ content of these pans reaches 40 to 50%. Usually there are two such pans facing each other with a root mat or decaying root material between them, indicating that their genesis is closely related to the plant growth.

Following GILE's et al. (1966) theory on the formation of carbonate horizons, the laminar horizon requires a plugged horizon to be formed. According to our observations as well as those of ROHDENBURG & SABELBERG (1979) this is not always the case. Indeed, the formation of the laminar shells may be one of the initial stages in the formation of a thick calcrete in some instances.

3.2. CARBONATISATION IN RELATION TO OTHER WEATHERING PROCESSES

The amount of carbonatisation in the subsoil is related to the release through weathering and the subsequent leaching of bases from the top soil. However, the various elements behave in different ways.

The differences in the elemental contents within the profiles IVB are too small to be interpreted as losses or gains in the soil.

The older soils (IVA and III) show a clear loss of calcium (compare Figure 2 and Tab. 3). In the horizons of maximum weathering intensity, the non-carbonate Ca decreases to 65% in IVA and to less than 10% in III of that of the parent material. The losses of Mg and Na show a similar tendency, but are somewhat less, whereas the gains and losses of K are variable. The release of all three elements Ca, Mg and Na is considerably slower from the solid basalt (profile III-570) than from the pyroclastic fall deposits (profile III-510) in the whole soil basis. A further indication of the slower weathering rates in the solid basalt is the considerable stone

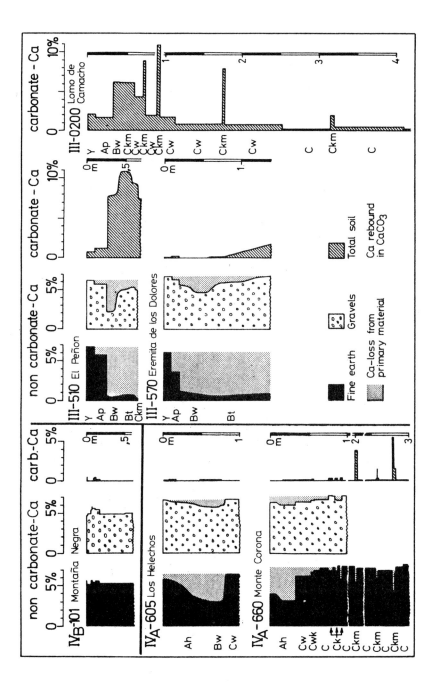

Fig. 2: Depth functions of non carbonate and carbonate calcium in soils of different age

content in profile III-570 compared to the Bt-orizon III-510, which is almost free of stones.

At least on surface III, Mg is bound as a carbonate. In profile III-510 a dissolution of the carbonates with HCl to pH 4 brought Mg into solution. If all this Mg was bound as a carbonate it would account for from 9% of the total carbonate in the Bw- to 17% in the Cwkm-horizons. X-ray analysis did not show any dolomite but revealed a shift of the main calcite peak to higher 2θ angles, indicating the presence of a magnesium bearing calcite. Measuring the shift, showed that 3 mol % $MgCO_3$ in the Bw- and 7% in the Cwkm-horizon is bound in the calcite.

An attempt was made to calculate the actual amount of Ca that has been released through weathering as well as the actual amounts rebound in $CaCO_3$.The main difficulties with these calculations are due to problems in measuring the bulk densities of the pyroclastic fall deposit and of the thin carbonate accumulation horizons.

The great ranges of released- and carbonate-bound Ca in Profile 660 originate from the uneven distribution of the soil moisture and, hence, differentiation in weathering intensities. The carbonate accumulation horizons, which themselves are variable in thickness and $CaCO_3$ content, also lead to horizontal movement of water and solutions, which makes representative sampling difficult.

In spite of these difficulties it may be assumed that only about 1/3 of the Ca released in profile IVA-660 is found bound in $CaCO_3$.

In landscape III, profile 510, the carbonate bound Ca reaches similar values as the released Ca, which we, however, explain by recent erosion and redeposition of the lower parts and at the bottom of slopes. The profile III-570, which due to an even stone distribution, we consider more or less undisturbed, contains only about 2 kg Ca/m^2 (0-100 cm) bound in carbonate. The profile III-0200, which was sampled to 4,5 m depth contains about 56 kg/m^2 bound as carbonate at depths ranging from 0,4 to 4,5 m. In this zone all Ckm-horizons occur.

Assuming that from the C-horizons in landscape III more Ca has been released than from the comparable horizons in IVA, we also have more Ca released through weathering in landscape III than has been rebound in carbonate.

In profile IVA-660 there is an obvious release of Ca from the C-horizons, in particular from the horizons with carbonate accumulation (compare Figure 2). Besides the carbonate accumulation these Ck-horizons are characterized by more intensive weathering and a finer texture caused by weathering and illuviation than the horizons above and below. Therefore,

Tab. 3: CALCIUM–RELEASE AND CARBONATE–CALCIUM IN SOILS OF DIFFERENT AGE

Profile and depth range	cm	Calcium (total soil)	
		released from primary minerals kg Ca/m^2	bound in $CaCO_3$ kg Ca/m^2
IVA -605	0-100	12	0.5
-660	0-100	12-16	2-4
	100-340	2-8	3-6
III -510	0-80	43	64
-570	0-100	65	2
	100-140	40	7
-0200	40-450	n.d.	56

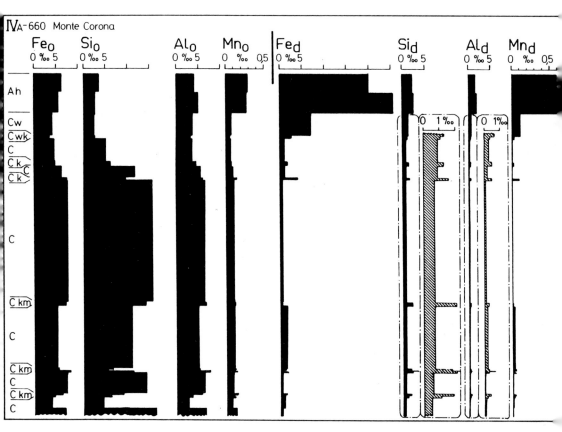

Fig. 3: Depth functions of ammonium-oxalate soluble (Fe_0, Si_0, Al_0 and Mn_0) and dithionite-citrate soluble (Fe_d, Si_d, Al_d and Mn_d) oxides in the soil IVA-660.

they may even show a greater release of Ca than is bound in carbonate.

The higher weathering intensities in the Ck-horizons are also apparent in the lower oxalate/dithionite-quotients compared to the horizons above and below. In particular, Si_0/Si_d and Al_0/Al_d are low whereas the Fe and Mn quotients may also be relatively high (Figures 3 and 4).

A comparison of the oxalate and dithionite soluble contents of the most intensively weathered, the least weathered and the Ck-horizons in profile IVA-660 is shown in Table 4.

These results indicate the more intensive weathering in the Ck- than in the other C-horizons and that the weathering is still in the "amorphous" phase as in the Ah-horizon of this mollic Andosol.

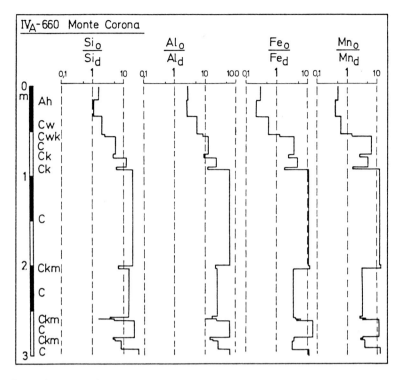

Fig. 4: Depth functions of the Fe, Si, Al and Mn ammonium-oxalate/dithionite-citrate quotients in the soil IVA.660.

Tab. 4: A COMPARISON OF THE WEATHERING INTENSITIES IN PROFILE IVA-660 USING CLAY CONTENT AND THE AMMONIUM–OXALATE AND DITHIONITE–CITRATE EXTRACTABLE ELEMENTS

Horizons		C		Ck and Ckm		Ah	
Nr. of horizons		8		10		2	
Fe_0	Means $^0/_{00}$	8,4	(4,0-14,1)	5,6	(3,5-8,1)	6,0	(5,6-6,3)
Si_0	(and $^0/_{00}$	17,7	(4,5-33)	7,6	(2,7-14,2)	3,3	(2,8-3,7)
Al_0	ranges) $^0/_{00}$	7,1	(2,6-12,4)	5,2	(3,3-7,4)	4,5	(4,0-5,0)
Mn_0	$^0/_{00}$	0,12	(0,07-0,18)	0,11	(0,06-0,16)	0,3	
Fe_d	$^0/_{00}$	0,7	(0,5-1,2)	2,5	(0,6-2,7)	23	(20-26)
Si_d	$^0/_{00}$	0,6	(0,5-0,7)	1,6	(1,1-2,1)	2,3	
Al_d	$^0/_{00}$	0,1	(<0,2)	0,4	(0,2-0,7)	1,7	(1,5-2,0)
Mn_d	$^0/_{00}$	0,01	(<0,02)	0,05	(0,01-0,13)	0,7	(0,6-0,8)
Clay content of fine earth	%	<1		2-13		16	
Texture:		sandy cinders to cindery coarse sand		cindery coarse sand to cindery sandy loam		cindery sandy loam	

4. HYPOTHESIS ON THE FORMATION OF THE CARBONATE ENRICHED HORIZONS

The macro- and micromorphological as well as the chemical analyses show that the process of carbonatisation is related to a number of other processes. The results have led to the following hypothesis on the formation of the Ck-horizons:
– The soil becomes moistened at regular intervals and occasionally water percolates through the soil.
– Calcium carbonate and fine earth accumulate at textural and bulk density bounaries. Carbonate coatings form on the Lapilli.
– The water holding capacity of the layers with the carbonate and fine earth accumulation is increased and they become invaded by plant roots. In the vertical root channels percolation rates increase. The channels become filled with fine earth and carbonates accumulate.
– The higher water holding capacity as well as the biological activity lead to intensification in weathering and hence to Ca release as well as to new formation of oxides and clay.
– The increased biological activity leads to higher CO_2-concentration in the soil solution. Water is removed through evapotranspiration and $CaCO_3$ precipitates, frequently at the root surface. Thus a root mat in the Ck-horizon may lead to the formation of the very thin and hard laminar shells in the center of the horizons.
– The intensive weathering in the top soil leads to high clay content and high water holding capacity. Thus a number of thin Ck-horizons may merge to form a major Ckm- or Bkm-horizon.

5. CONCLUSIONS

The amounts of Ca released through weathering are in all the soils higher than the Ca bound in the carbonates, the relation being about 3 to 1. The Ca released through weathering is in IVA mainly from the top meter of the profile, and in III from the top 1,5 meter.

The Ca accumulates in numerous thin horizons from about 1 m to about 4,5 m depth. In these thin horizons the contents of organic matter, clay and dithionite soluble Fe, Al, Si and Mn are higher than in the C-horizons above and below, but lower than in the Ah-horizons. Thus the weathering intensity increases from the C- over Ck- to Ah-horizons.

A hypothesis on the formation of the Ck-horizons is postulated on the basis of the field observation, micromorphology and chemical analyses. It states that the initial stage in the calcrete formation are the thin carbonate accumulation horizons and that these merge near the surface to form a major Ckm- or Bkm-horizon.

ACKNOWLEDGEMENTS

This research was financed through a grant from the Deutsche Forschungsgemeinschaft. The authors want to thank Frau U. Förster and Frau S. Juenge for helping with the chemical analyses.

REFERENCES

FITZPATRICK, E.A. (1980): The micromorphology of soils, A manual for the preparation and description of thin sections of soils. Department of Soil Science, University of Aberdeen.
FUSTER, J.M.S., SANTIN FERNANDEZ S. & SAGREDO, J. (1968): Geologia Y Volcanologia de las Islas Canarias, Lanzarote. Inst. "Lucas Mallada" C.S.I.C., Madrid.
GILE, L.H., PETERSON, F.F. & GROSSMAN, R.B. (1966): Morphological and genetic sequence of carbonate accumulation in desert soils. Soil Science 101, 347-360.
GUDMUNDSSON, Th., JAHN, R. & STAHR, K. (1984): Kalkanreicherung und Krustenbildung in Böden aus jungen Vulkaniten Lanzarotes. Mitteilungen Deutsche Bodenkundliche Gesellschaft 39, 25-28.
INSTITUTO GEOLOGICO Y MINERO DE ESPANA (1967-1968) (Edit.): Mapa Escala 1:50 000 (8 hojas con memorias). Inst. "Lucas Mallada", C.S.I.C., Madrid.
JAHN, R., STAHR, K. & GUDMUNDSSON, Th. (1983): Bodenentwicklung aus tertiären bis holozänen Vulkaniten im semiariden Klima Lanzarotes (Kanarische Inseln). Z. Geomorph. N.F., Suppl.-Bd. 48, 117-129.
ROHDENBURG, H. & SABELBERG, U. (1979): Kalkkrustentypen im westlichen Mediterrangebiet (Spanien und Marokko). Landschaftsgenese und Landschaftsökologie 5, 25-32. Lehrstuhl für Physische Geographie und Landschaftsökologie der TU Braunschweig.
SCHLICHTING, E. & BLUME, H.P. (1966): Bodenkundliches Praktikum. Paul Parey, Hamburg und Berlin.
WIEDER, M. & YAALON, D.H. (1982): Micromorphological fabrics and development stages of carbonate nodular forms related to soil characteristics. Geoderma 28, 203-220.

Address of authors:
Reinhold Jahn, Thorsteinn Gudmundsson und Karl Stahr
Department of Ecology – Soil Geography, Technische Universität Berlin
Salzufer 12, D 1000 Berlin 10, Germany

E. Fernandez Caldas & Dan H. Yaalon (Eds):
VOLCANIC SOILS
CATENA SUPPLEMENT 7, Braunschweig 1985

CHARACTERISTICS OF THE VANUATU ANDOSOLS

P. **Quantin,** Bondy

SUMMARY

The andosols of the Vanuatu Archipelago are deriving from recent volcanic ashes and lapilli, under a tropical climate. There are four main groups of andosols: vitric (Vitrandepts), base-saturated (Eutrandepts), base-unaturated (Dystrandepts) and perhydrated (Hydrandepts).

This paper summarizes the main characteristics of each group and it emphasizes their diagnostic criteria. These are: for the first two groups, the richness in silica of the weathering products and a rather weak or moderate water and phosphorus retention-capacity: for the last two groups, the richness in alumina of the weathering products and a high water and phosphorus retention-capacity. Other diagnostic criteria have been selected, like: the dehydration-irreversibility and $\triangle$ CEC ratios, the bulk density, the Al-oxalate content and the base-saturation ratio. Among them the first two are the most note-worthy.

RESUME

Les andosols de l'Archipel du Vanuatu sont dérivés de cendres et lapilli volcaniques récents, en climat tropical. Il y en a quatre groupes principaux: vitriques (Vitrandepts), saturés en bases (Eutrandepts), désaturés en bases (Dystrandepts) et perhydratés (Hydrandepts).

Cet article résume les caractéristiques majeures de chaque groupe et il met en relief leurs critères diagnostics. Ce sont: pour les deux premiers, la richesse en silice des produits d'altération et une capacité de rétention pour l'eau et le phosphore plutôt faible ou modérée; pour les deux autres groupes, la richesse en alumine des produits d'altération et une capacité élevée de rétention pour l'eau et le phosphore. D'autres critères diagnostics ont été sélectionnés, tels que: les taux de déshydratation irréversible et de $\triangle$ CEC, la densité apparente, l'extrait d'Al-oxalate et le taux de saturation en bases, parmi lesquels les deux premiers sont les plus remarquables.

RESUMEN

Los andosols del Archipiélago de Vanuatu provienen de cenizas y escorias volcánicas recientes, en un ambiente tropical. Se distinguen cuatro groupos principales: vítricos (Vitrandepts), saturados en bases cambiables (Eutrandepts), desaturados en bases cambiables (Distrandepts) y perihidratados (Hidrandepts).

En este papel se compendian las caracteristicas mayores de cada grupo y se hacen resaltar los criterios de diagnóstico de esos. Conviene destacar: en los dos primeros grupos, la riqueza en silice de los productos de la meteorozación y una capacidad bastante debil o moderada de retención del agua y del fósforo; en los dos otros grupos, la riqueza en alumina de los productos de la mateorización y una fuerte capacidad de retención del agua y del fósforo. Otros criterios de diagnóstico han sido seleccionados, tal como: los porcentajes de deshidratación irreversibile, de $\triangle$ CEC y de saturación en bases cambiables, la densidad aparente y el extracto de Al-oxalato, siendo más relevantes los dos primeros criterios.

ISSN 0722-0723 / ISBN 3-923381-06-9

1. INTRODUCTION

The andosols of the Vanuatu are located in a tropical country, between 13° and 20° S Latitude, in the S.W. Pacific. They are found under a warm and wet climate. These soils have been deriving from recent (< 5,000 years) volcaniclastic materials, such as ash and lapilli. These materials are mainly basic and of basalt and andesite composition. We have observed the following four main groups of Andosols: 1. undifferentiated and vitric (Vitrandepts); 2. well differentiated and base-saturated (Eutrandepts); well differentiated, base-unsaturated and unperhydrated (Dystrandepts); well differenciated, base-unsaturated and perhydrated (Hydrandepts).

2. CHARACTERISTICS

The main characteristics of the four groups of Andosols have been summarized in the tables 1a, 1b, 1c.

Tab. 1a: MAIN CHARACTERISTICS OF THE 4 GROUPS OF ANDOSOLS (IN THE B HORIZON FOR THE GROUPS II, III AND IV).

Andosol-group	I Vitric	II eutric	III perhydric	IV dystric
Age of material / Years	< 1,000	1,000 - 2,000	2,000 - 5,000	2,000 - 5,000
Rainfall mm/y	> 2,000	1,800 - 4,000	> 4,000	2,500 - 4,000
Presence of (B) horizon	No	Yes	Yes	Yes
Bulk density g/cc	1.2 - 0.9	0.9 - 0.7	0.3	0.5 - 0.4
Primary minerals %	> 60	30 - 60	10 - 50	10 - 50
Organic matter A_1 %	> 1 / < 10	5 - 10	> 20	10 - 20
(B) %	–	1 - 5	2 - 4	1 - 4
< 2 μ fraction %	< 10	10 - 30	20 - 60	> 20

Tab. 1b: MAIN CHARACTERISTICS OF THE 4 GROUPS OF ANDOSOLS (IN THE B HORIZON FOR THE GROUPS II, III AND IV).

Andosol-group		I Vitric	II eutric	III perhydric	IV dystric
Secondary minerals		opal Si-allophane	Fe-allophane	Imogolite Gibbsite	Al-allophane
Amorphous products	Seg. %	< 15	11 - 24	25 - 55	> 20
(Al, Fe, Si) amount*	Oxal. %	< 8	6 - 13	15 - 39	> 15
	HCl %	< 5	5 - 10	18 - 31	> 10
Al_2O_3 oxalate** %		0.4 - 2.8	a: 1 - 5 b: 0.2 - 1	6 - 19	> 5
Allophanic clays %		< 5	5 - 10	15 - 30	> 10
SiO_2/Al_2O_3 mol. of < 2 μ fraction		> 3	> 2	≤ 1	1 - 2
Specific surface area m²/g of < 2 μ fraction		100 - 200	100 - 200	300 - 700	300 - 400

*   Sum of amorphous Al, Fe and Si oxides extracted by Seg = Segalen method, Oxal = Tam reagent
      with U.V., HCl = 2 N HCl reagent.
** Al_2O_3 extracted by NH_4-oxalate pH 3.5 with U.V., infered from cumulative curve of 6 extracts.
      a: chromic subgroup; b: melanic subgroup.

Tab. 1c: MAIN CHARACTERISTICS OF THE 4 GROUPS OF ANDOSOLS (IN THE B HORIZON
FOR THE GROUPS II, III and IV).

Andosol-group	I Vitric	II eutric	III perhydric	IV dystric
pH (H_2O)	~ 7	~ 7	~ 6	~ 6
$\triangle$pH (H_2O - KCl)	–0.9 to –0.5	–1.0 to –0.5	–0.4 to + 0.2	–0.5 to –0.3
CEC (pH 7) me/100 g	< 15	20 - 40	10 - 30	15 - 25
$\triangle$CEC (pH 9 - pH 4) ratio***	> 40	40 - 60	70 - 80	60 - 70
Base-Saturation ratio %	> 50	> 50	$\leqslant$ 10	20 - 50
Water-retention-capacity % ⅓b	< 45	40 - 85	100 - 300	80 - 130
(wet soil) 15b	< 25	20 - 60	80 - 240	50 - 100
Dehydration-irreversibility ratio**** %	> 40	40 - 60	75 - 90	50 - 73
P retention-capacity %	a: 0 - 8 b: 25 - 55	c: 65 - 85 d: 40 - 50	97 - 99	98 - 99
available P (Truog) ppm	$\geqslant$ 15	$\geqslant$ 50	0	0

*** $\dfrac{\triangle \text{CEC (pH 9 - pH 4)}}{\text{CEC (pH 9)}} \times 100$

**** $\dfrac{\triangle H_2O \text{ (⅓ b) wet - air dried sample}}{H_2O \text{ (⅓ b) wet sample}} \times 100$

2.1.   VITRIC ANDOSOLS

The "vitric andosols" are very young soils which derive from less 1,000 years old volcaniclastic material. The weathering is slightly advanced and the soil profile is rather undifferentiated, although several layers of ash and lapilli are often superimposed. These soils consist
essentially of one or several shallow humic horizons, containing 1-8% organic matter and 1-
10% of < 2 μ fraction.

The vitric andosols contain more than 60% of unweathered pyroclastic material, such as
glass and microlites. Their weathering products are not made of clay minerals, except some
eventual traces of halloysite and smectite, but only of amorphous or paracrystalline constituents, like allophane and opal. These products are very rich in silica, but very poor in alumina (mol. SiO_2/Al_2O_3 ratio > 3). Then, the silica and iron-rich form of allophane seems
similar to hisingerite. The amount of estimated amorphous products is at most: 15 % by
Segalen's method, 8% by oxalate extraction, or 5% by 2N HCl dissolution. In fact, the
allophane content does not exceed 5% of whole soil.

The vitric andosols have the following main characteristics: a bulk density ~ 0.9 - 1.2;
a light or even null reaction to the NaF test; a water-retention capacity (undried soil) < 45%
at ⅓ bar and < 25% at 15 bars; a rate of dehydration-irreversibility after air-drying > 50%; a
pH weakly acid in the topsoil (except in case of sulfuric fumaroles or very heavy rainfall), neutral or slightly alkaline in the depth; a CEC value < 25 me in A_1 horizon, or < 15 me/100 g
in C horizons, but CEC/< 2 μ fraction > > 1 me/1 g; a rate of $\triangle$ CEC 40%; a base-
saturation ratio > 50% (except the soils under a perhumid climate or to windward of sulfuric
fumaroles); available phosphorus (Truog's method) values fairly high > > 15 ppm, which
means a weak P-retention (25-55%).

These soils can be classified as Vitrandepts in the Soil Taxonomy (1975), or diverse vitric great groups or subgroups of Andisol Proposal (1978).

2.2. SATURATED ANDOSOLS

The "saturated andosols" are young soils. Their parent-material dates from the recent past, $\sim$ 1,000 to 2,000 years ago. Its chemical composition is rather basic and rich in iron. These andosols are forming under a wet tropical climate, which comprises or not a short dry season, but is never perhumid (rainfall < 3,000 mm/y). Their profile is fairly well differentiated in A_1, (B) or B-C, and C horizons. But the (B) horizon is at an early stage of development (1-2 m) and the weathering is weakly advanced. Often, the soil is multiphase and made of a serie of successive (often 2 or 3, sometimes as many as 7) eruption deposits.

The upper humic horizon is 10-25 cm thick and very dark in colour; it contains 5-10% organic matter and 10-30% of < 2 μ fraction. The (B) horizon is brown or reddish-brown coloured; it contains 1-5% organic matter and 5-30% of < 2 μ fraction.

The mineral soil is still rich in weatherable primary minerals, the amount of which approximates 30-40 %. The weathering products are largely amorphous or paracrystalline, at least in the upper part of soil profile; although small quantities of clay minerals, such as halloysite in the wettest regions or beidellite in the driest zones, appear increasing toward the deeper part of soil or toward the drier climatic zone. The weathering products are very rich in silica, iron and basic cations, but rather poor in alumina (mol. SiO_2/Al_2O_3 ratio generally > 2). These products are rich in opal, such as diatom skeletons, phytoliths and micro-discs. In addition they contain an iron-rich allophane which is similar to hiseringite. But they have few iron oxy-hydroxides and no free alumina. The amount of amorphous products in the (B) horizon approximates: 11-24% (as Al, Fe and Si oxides) by Segalen's method, 6-13% by oxalate extraction or 5-10% by 2N HCl dissolution. In fact, the allophane content does not exceed 10% of whole soil.

The saturated andosols have the following main properties in the (B) horizon: a bulk density $\sim$ 0.7 - 0.9; a fairly good reaction to the NaF test; a water-retention capacity (undried soil) $\sim$ 40-85 % at $\frac{1}{3}$ bar and $\sim$ 20-60 % at 15 bars; a rate of dehydration-irreversibility $\sim$ 40-60 %; a pH near 7 and $\triangle$ pH $\sim$ 0.5 to 1.0; a CEC value $\sim$ 20-40 me/100 g and CEC/ < 2 μ fraction > > 1 me/1 g; a rate of $\triangle$ CEC $\sim$ 40-60 %; a base-saturation ratio > 50 %; available phosphorus (Truog's method) > 50 ppm, meaning a weak or moderate P-retention, < 85 %).

These soils can be classified as Eutrandepts in the Soil Taxonomy (1975). But, their taxonomic position in the Andisol Proposal (1978 or 1983) is not very clear; a mollic subgroup should be proposed for the Haplotropands (1978) or Tropudands and Haplustands (1983) great groups. We proposed in our classification proposal (Groupe de Travail Andosol, 1972) to subdivide the saturated andosols in: a chromic subgroup for the reddish-brown ones of the wetter zone and a melanic subgroup for the dark epipedon type of the drier zone. This latter type has a tendency to form beidellite clay minerals and then to have smaller values of both $\triangle$ CEC and dehydration-irreversibility rates.

2.3. PERHYDRATED ANDOSOLS

The "perhydrated andosols" are unsaturated and very hydrated soils. They are still young soils, but at a more advanced stage of weathering than the vitric or saturated andosols. Their parent-material dates several thousand years; but it is generally less 5,000 years old and often rejuvenated by recent ash-falls in the topsoil. Moreover, the climate is perhumid (rainfall > 4000 mm/y). The soil profile is well differentiated in A_1 (B) and C horizons; it is often multiphase and shows a more weathered IIB horizon; its depth can exceed 2 m.

The upper humic horizon is 10-30 cm thick, very dark in colour, and contains often > 20 % of organic matter. The transition to B horizon is gradual, the organic matter decreasing to 15-10 % in the A-(B) horizon until 40-80 cm, and then 4-2 % in the B horizons between 1-2 m in depth. The B horizon has a fine loamy texture and shows a massive but microporous structure and some thixotropic properties.

The perhydrated andosols have undergone a very strong weathering, at least in the (B) or II B horizons, where the residue of primary minerals can approximate 30 to 10 %. The weathering products are almost amorphous or paracrystalline, even in the lower and older part of the soil. These nearly amorphous minerals are very abundant, approximating 15-30 % in the A and 30-60 % in the B horizons. Nevertheless, these can contain some traces of clay minerals (halloysite, kaolinite and even smectite), Fe-oxyhydroxides (goethite, lepidocrocite) and moreover a few gibbsite, which increases often toward te bottom, in the older part of soil. These weathering products are very rich in alumina; their molar SiO_2/Al_2O_3 ratio approximates 1 and it is often lesser. They are mainly constituted by imogolite (20 - 25 % of soil), with some allophane and Al-Fe hydroxide gels. In the top-soil the allophane increases and few opal appears, due to the enrichment of soil in fresh minerals and silica by new ash-falls. These products have a very large specific surface area: $\sim$ 300 - 700 m^2/g.

The perhydrated andosols show the following main properties in the B horizons: a bulk density $\sim$ 0.3; a very strong reaction to the NaF test; a water-retention capacity (undried soil) generally much more than 100 %, $\sim$ 100 - 300 % at $\frac{1}{3}$ bar, $\sim$ 80-240 % at 15 bars, and moreover a rate of dehydration-irreversibility $\sim$ 75-90%; a pH $\sim$ 6 and $\triangle$pH $\sim$ -0.4 to + 0.2; a CEC value $\sim$ 10-30 me/100 g, CEC / < 2 μ fraction generally near 1 ($\sim$ 0.3-1.4) me/1 g; moreover a rate of $\triangle$ CEC $\sim$ 70 - 80 %; a base-saturation ratio often < 10 %; available phosphorus (Truog's method) $\sim$ 0 ppm, meaning a very high P-retention (> 95 %).

Most of these soils could be classified as Hydrandepts in the Soil Taxonomy (1975), or Hydrudands for Andisol Proposal (1978).

2.4. UNSATURATED ANDOSOLS

The "unsaturated andosols" are the more common andosols of lowlands, which are under a wett but non-perhumid climate (rainfall $\sim$ 2500-4000 mm). They form an intergrade between the both preceding groups of andosols. Like the perhydrated ones, they are young soils, dating several thousand years, showing a well differentiated profile and having a low base-saturation in the B horizon. But according to less heavy rainfall conditions, they are not at a so advanced stage of weathering and their base-saturation ratio is higher, $\sim$ 20-50 %. In addition, the B horizon has a more granular, although very friable, structure, and doesn't show any thixotropic property.

The upper humic horizon is more reddish and contains only 10-20 % organic matter, while the B horizon keeps still 1-4 % of it. But, the amount of < 2 μ fraction and the residue of weatherable minerals don't differ greatly. Nevertheless the content in amorphous or paracrystalline products is lesser than in perhydrated andosols and moreover, these products are richer in silica, the molar SiO_2/Al_2O_3 ratio being $\sim$ 1-2. There is a mixture of sphaeroidal allophane and fibrous imogolite. But in the depth, the oldest part of soils shows often a few quantity of halloysite, goethite and gibbsite. However, the specific surface area of $< 2\,\mu$ fraction from the B horizon remains high: $\sim$ 300 - 400 m^2/g.

The unsaturated (non-perhydrated) andosols present the following main properties in the B horizon: a bulk density $\sim$ 0.4-0.5; a quite good reaction to the NaF test; a water retention capacity (undried soil) $\sim$ 80-130 % at $\frac{1}{3}$ bar and 50-100 % at 15 bars; a rate of dehydration irreversibility $\sim$ 50-73 %; a pH near 6 (5.7 - 6.7) and $\triangle$pH $\sim$ 0.5 to 0.3; a CEC value $\sim$ 15-25 me/100 g and CEC/ $< 2\,\mu$ fraction often < 1 ($\sim$ 0.4 to 1.4); a rate of $\triangle$ CEC $\sim$ 60-70 %; a base-saturation ratio $\sim$ 20-50 %; available phosphorus (Truog's method) $\sim$ 0 ppm and a high P-retention ($>$ 95 %).

These soils can be classified as Dystrandepts in the Soil Taxonomy (1975), and Haplotropands or Tropudands for Andisol Proposal (1978, 1983).

3. CONCLUSION

We have seen that the Vanuatu andosols can be subdivided in four main groups. These andosols derive from recent and basic volcaniclastic materials. Both main factors of differentiation are the age of ashfall parent material and the rainfall distribution.

The "vitric" ones are the less 1,000 years old andosols, without B horizon. These soils show a very weak grade of weathering, and have the following properties: unweathered primary material $>$ 60 %; bulk-density $\sim$ 1.2 - 0.9; silica-rich weathering products, made of opal and iron-rich allophane; molar SiO_2/Al_2O_3 ratio of $< 2\,\mu$ fraction $>$ 3; water-retention $>$ 3; water-retention capacity* at $\frac{1}{3}$ bar < 40 %, at 15 bars < 20 % and rate of dehydration-irreversibility** > 40 %; CEC < 15 me 100 g, rate of $\triangle$ CEC*** > 40 % and base-saturation ratio**** generally > 50 %; low P-retention (< 55 %).

The "saturated" andosols date around 1,000 - 2,000 years and show a beginning of (B) horizon, although they are still weakly weathered. They don't appear under a perhumid climate. There are their main properties in the (B) horizon: unweathered primary material < 60 %; bulk density $\sim$ 0.9-0.7; silica and iron-rich weathering products, made of hisingerite-like allophane and few opal; molar SiO_2/Al_2O_3 ratio of $< 2\,\mu$ fraction > 2; water-retention capacity * at $\frac{1}{3}$ bar $\sim$ 40-80 %, at 15 bars $\sim$ 20-60 %, and dehydration-irreversibility rate** $\sim$ 40-60 %; CEC $\sim$ 20-40 me/100 g, rate of $\triangle$ CEC*** $\sim$ 40-60 % and base-saturation ratio**** > 50 %; low P-retention (< 85 %).

The "perhydrated" andosols date several thousand years and show a well developed B horizon and the most advanced grade of weathering. They form only under a perhumid climate (> 4000 mm/y rainfall). The upper A_1 horizon contains at least 20 % organic matter. The B horizons show the following properties: unweathered primary material < 50 %; bulk-density $\sim$ 0.3; > 30 % of alumina-rich weathering products, made of imogolite, $\pm$ allophane and few gibbsite; molar SiO_2/Al_2O_3 of $< 2\,\mu$ fraction < 1; water-retention capacity* at $\frac{1}{3}$ bar $\sim$ 100-300 %, at 15 bars $\sim$ 80-240 %, and dehydration-irreversibility rate** $\sim$ 75-90 %; rate of $\triangle$ CEC*** $\sim$ 70-80 %; base-saturation ratio**** $\sim$ 10 % or less; very high P-retention (> 95 %).

---

*   on undried soil
**   W.R.C. at $\frac{1}{3}$ bar
***   rate of pH dependant CEC between pH 9 and 4
**** base-saturation of CEC at pH 7

The "unsaturated" andosols are close the perhydrated ones, but they are not formed under a perhumid climate (rainfall $\sim$ 2500-4000 mm). The A_1 horizon contains only 10-20 % organic matter. In the B horizons their properties differ by the following ranges of value: bulk-density $\sim$ 0.5-0.4; molar SiO_2/Al_2O_3 ratio of $< 2 \mu$ fraction $\sim$ 1-2; paracrystalline clay richer in allophane than in imogolite; water-retention capacity* at $^1/_3$ bar $\sim$ 80-130 %, at 15 bars $\sim$ 50-100 % and moreover a rate of dehydration irreversibility** $\sim$ 50-73 %; rate of $\triangle$ CEC*** $\sim$ 60-70 %; base-saturation ratio**** $\sim$ 20-50 %. The P-retention remains high ($>$ 95 %).

In conclusion we can outline these following criteria as diagnostic for the four group of Vanuatu andosols: presence or not of B horizon and several ranges of bulk density; silica richness of weathering products, marked by the molar SiO_2/Al_2O_3 ratio of $< 2 \mu$ fraction; water-retention capacity at 15 bars (on undried soil) and rate of dehydration irreversibility at $^1/_3$ bar; CEC and moreover the rate of $\triangle$ CEC; base-saturation ratio; and P-retention capacity. These criteria have been summarized in the table 2.

Tab. 2: DIAGNOSTIC CRITERIA OF THE FOUR GROUPS OF ANDOSOLS (IN THE B HORIZON FOR THE GROUPS II, III AND IV).

Andosol group	I Vitric	II eutric	III perhydric	IV dystric
Age y.	$<$ 1,000	1,000 - 2,000	$>$ 2,000	$>$ 2,000
Rainfall mm/y	$\sim$	$<$ 4,000	$>$ 4,000	2,500 - 4,000
(B) horizon	No	$+$	$+$	$+$
Bulk-density	1.2 - 0.9	0.9 - 0.7	0.3	0.5 - 0.4
SiO_2/Al_2O_3 mol. of $< 2 \mu$ fraction	$>$ 3	$>$ 2	$\leqslant$ 1	1 - 2
H_2O 15 b. (wet soil) %	$<$ 20	20 - 60	80 - 240	50 - 100
Dehydration-irreversibility-ratio %	$>$ 40	40 - 60	75 - 90	50 - 75
$\triangle$ CEC ratio %	$>$ 40	40 - 60	70 - 80	60 - 70
Base-saturation %	$>$ 50	$\geqslant$ 50	$\leqslant$ 10	20 - 50
P retention-capacity %	$<$ 60	40 - 85	$>$ 95	$>$ 95
Al_2O_3-oxalate %	$<$ 2	$<$ 5	$>$ 5	$>$ 5

BIBLIOGRAPHY

GROUPE DE TRAVAIL ANDOSOLS (CPCS) (1972): Cahiers ORSTOM, sér. Pédologie, X (3), 302-303.
ICOMAND (1983): Circular letter n. 5. New Zealand Soil Bureau, Lower Hutt, N.Z.
QUANTIN, P. (1972-1978): Atlas des Nouvelles-Hébrides. ORSTOM, Paris.
QUANTIN, P. (1982): Vanuatu Agronomic Potential and Land Use Map. ORSTOM, Paris.
QUANTIN, P. (1982): Cahiers ORSTOM, sér. Pédologie, XIX (4), 369-380.
SMITH, G. (1978): The Andisol Proposal. New Zealand Soil Bureau, Lower Hutt, N.Z.
U.S.D.A. (1975): Soil Taxonomy. Soil Conservation Service, Washington.

Address of author:
P. Quantin, ORSTOM, 70, route d'Aulnay
93140 Bondy, France

E. Fernandez Caldas & Dan H. Yaalon (Eds):
VOLCANIC SOILS
CATENA SUPPLEMENT 7, Braunschweig 1985

CHARACTERISTICS AND GENESIS
OF TWO ANDOSOLS IN CENTRAL ITALY

P. **Quantin,** B. **Dabin** and A. **Bouleau,** Bondy
L. **Lulli** and D. **Bidini,** Firenze

SUMMARY

Two types of andosols were observed near Lake Vico (Latium), on an extinct volcano. Both derived from ashes and lapilli of the Late-Pleistocene, but the climatic conditions were different.

The first one is located on Mt. Fogliano (940 m o.s.l.), under the coldest and wettest climate of this region and under a beech-forest. It is a weakly differentiated andosol, of a humic and melanic type (Melanalland). This soil contains aluminium-rich amorphous organo-mineral products, which have a high ratio of variable charges ($\sim$ 80%). The pH is $\sim$ 5-6, in spite of a base-saturation ratio $\sim$ 20%. This andosol shows a high content in chelated-Al in the top-soil, followed by the formation of proto-imogolite in the middle of the profile and the sudden appearance of halloysite at its bottom, in coarser lapilli.

The second soil is located on the South-slope of Mt. Venere (700 m o.s.l.), under a sunnier climate and an oak-forest. It is a well differentiated andosol, of a moderately unsaturated and chromic type (Hapludand). This soil does not show chelated-Al; its pH is higher and the rate of variable charges is less ($\sim$ 55%) than in the former. Some allophane is forming in the top of the profile, while halloysite appears at its bottom, in scoriaceous lapilli.

Some hypothesis are suggested on the genesis of both types of andosols, in a mediterranean country, according to the peculiarities of their climate conditions and of their organic constituents.

RESUME

Deux types d'andosols ont été observés près du Lac Vico (Latium), sur un volcan éteint. Les deux dérivent de cendres et lapilli du Pléistocène supérieur. Mais les conditions climatiques diffèrent.

Le premier est situé sur le Mt. Fogliano (alt. 940 m), sous le climat le plus froid et humide de la région et sous une forêt de hêtres. C'est un andosol faiblement différencié, de type humique et mélanique. Il est riche en produits organo-minéraux amorphes et très alumineux, à fort taux de charges variables ($\sim$ 80 %). Le pH est $\sim$ 5-6 magré un taux de saturation en bases de $\sim$ 20%. Ce sol présente une forte teneur en Al-chélaté en haut du profil, la formation de proto-imogolite dans sa partie médiane et l'apparition brusque d'halloysite à sa base, dans des lapilli plus grossiers.

Le deuxième est situé sur le versant Sud du Mt. Venere (alt. 700 m), sous un climat plus ensoleillé et une forêt de chênes. C'est un andosol bien différencié, de type modérément désaturé et chromique. Il ne présente plus d'Al-chélaté; le pH est moins acide et le taux de charges variables diminue ($\sim$ 55 %). Il se forme de l'allophane dans le haut du profil et de l'halloysite à sa base, dans des lapilli scoriacés.

Des hypothèses sont proposées sur la genèse de ces deux types d'andosol, en région méditerranéenne, en relation avec les particularités de leurs conditions climatiques et de leurs constituants organiques.

ISSN 0722–0723 / ISBN 3–923381–06–9

RESUMEN

Dos tipos de andosol han sido observados cerca de Lago Vico (Latium), sobre un volcán apagado. Ambos derivan de cenizas y escorias del Pleistocéno superior, pero bajo dos condiciones climáticas differentes.

El primero se encuentra sobre el Mte Fogliano (940 m s.n.m.), con el clima más frio y humedo de la región y bajo un hayal. Este suelo es ríco en productos organo-minerales amorfos y muy aluminosos, con una fuerte proporción de cargas variables ($\sim$ 80 %). El pH es $\sim$ 5-6, mientras el porcentaje de saturación en bases cambiales alcanza $\sim$ 20 %. Este suelo muestra un fuerte contenido de Al-chelatado en la parte superior, mientras la protoimogolita se forma en el medio del perfil y la haloisita aparece subitamente abajo, en las escorias más gruesas.

El secundo suelo está ubicado sobre el vertiente sur del Mte. Venere (700 m s.n.m.), con un clima más asoleado y bajo un encinar Se trata de un andosol bien diferenciado, de tipo moderadamente desaturado y cromico. No tiene más Al-chelatado y la proporción de cargas variables decrece ($\sim$ 55 %). Alofana está formandose en la parte superior del perfil y haloisita abajo, en los lapilli escoriáceos.

Unas hipótesis son propuestas sobre la génesis de ambos tipos de andosol, en una región mediterránea, en conexion con las particularidades climaticas y los constituyentes organicos de ellos.

1. INTRODUCTION

Some andosols have been observed at the top of the Vico volcano, in Central Italy. They are lying on the most recent pyroclastic products and under the most temperate and wet climate of this region: either outside the caldera at its rim, on Mte Fogliano, or inside on the summit of Mte Venere cone. Two types of andosols are distinguished. One, on Mte Fogliano, has a very deep humic horizon, without B horizon; other, on Mte Venere, has a well differentiated B horizon. We will show the main characteristics of the both soils, and then deduce their meaning about the genesis and the classification.

2. LOCATION, ENVIRONMENT

Vico is located in the North of Roma, at 42°20'N and 12°10' O. The volcano forms a wide cone. But it is depressed by a central caldera, in the bottom of which (atl. 510 m) lies the Vico lake and in the middle rises the pyroclastic cone of Mte Venere (alt. 838 m). Mte Fogliano is the highest point (alt. 965 m) on the western flank of the caldera. The first andosol (profile 12) has been observed near the top of Mte Fogliano (alt. 940 m), the second (profile 13) on the southern slope of Mte Venere (alt. 700 m). On the Mte Fogliano, the climate is colder and cloudy, of "perhumid" type, and the vegetation is a beech-forest; on the western slope of Mte Venere, the climate is sunnier and the vegetation is an oak-forest. In the both situations the parent material originates from the last eruptions of Mte Venere, the age of which is recent (< 90.000 ans). There are alkaline-trachytic ashes in the upper part of soil and phonolitic-tephritic cinders at their bottom. These products are vitreous and very porous, rich in microlites and phenocrystals of K-feldspars, and then in aluminium and potassium but poor in calcium, iron and magnesium. In addition the drainage of the rain is very high in the soil.

3. MORPHOLOGY

3.1. MTE FOGLIANO SOIL (Profile 12)

The profile shows successively the following horizons: A_0 (0-5 cm), organic leaf-litter – A_1 (5-42 cm), black and very humiferous – A_{13} (42-83 cm), slightly more brownish and less humiferous – C_1 (83-100 cm), brownified ash – C_2 (100 to $>$ 120 cm), weathered cinder. Then this soil is little differentiated, black and very humiferous. The loamy texture of humic horizons, the very fine and friable structure, the low bulk-density (0.8-0.9) and a very high reaction to NaF test, characterize an andosol.

3.2. MTE VENERE SOIL (Profile 13)

The main horizons are successively: A_1 (0-48 cm), very dark brown and very humiferous – A_3 (48-68 cm), yellowish-brown, gradual transition – B_1 (68-96 cm), more reddish – B_2 (96-120 cm), more brownish, slightly/sticky and plastic, containing more cinders – B_3 (120 to $>$ 140 cm), brownish and weathered cinders. This soil is well differentiated. For the most part (0-98 cm), it has typical characteristics of andosol: a fine texture, a very friable structure, a low bulk density (0.9) and a high reaction to the NaF test. In the depth, the soil becomes slightly clayey and plastic, and it reacts more slowly to the NaF test. This shows a difference in the nature of the weathering products according to a change of the parent material.

4. MINERALOGY, GEOCHEMISTRY

4.1. MTE FOGLIANO SOIL

The mineral part of soil is constituted at least by 75 % of unweathered glass and of primary minerals (residue determined after acid dissolution). The latter comprise near 90 % of feldspars, $2/3$ of which are made of sanidine and $1/3$ of plagioclases (albite-oligoclase), and only 10 % of ferro-magnesian minerals, mainly aegyrinic-augite, plus a few biotite and basaltic hornblend, and traces of ilmenite (determined by X Ray diffraction and optical microscopy). The whole composition of these minerals is similar of an alkaline trachyte.

The weathered part of soil (extracted by ORSTOM Triacid-method) is very impoverished in silica and relatively enriched in aluminium and iron. The amorphous and paracrystalline (selectively dissolved by 2N HCl, or oxalate) fraction, constitutes only 6 to 11 % of soil, 3 to 5 % of which are allophane (imogolite-formula), and 3 to 6 % are chelated or para-crystalline Al and Fe hydroxides. The molar SiO_2/Al_2O_3 ratio near 0.5 shows these products are very aluminous. The $< 2 \mu$ fraction is constituted mainly of gels of fibrous allophane (observed by TEM), that looks like the proto-imogolite (FARMER et al. 1978, VIOLANTE & TAIT 1979), as also of organo-mineral complexes. For the most part of soil, there are only traces of clay minerals: (10 Å) halloysite and (10-14 M) altered mica. However at the bottom of soil profile, in the brownish cinders, sphaeroidal (10 Å) halloysite appears suddenly in a great quantity (> 10 %), while allophane decreases strongly (to 2 %). The quantity of allophane increases first until the bottom of humic horizons and after it decreases in the mineral horizon to the benefit of halloysite.

8

4.2. MTE VENERE SOIL

The mineral composition is similar to that of Mte Fogliano soil. However, there are some significant differences. The part of unaltered glasses and phenocrystals is restricted to 65-60 % in the A and B_1 horizons, even 55 % in the weathered cinders, at the soil bottom. The weathering seems more advanced than in the Mte Fogliano soil. In addition, the proportion of plagioclase is higher (~ 50 % of feldspars). The volcanic parent material is a little richer in iron, calcium and magnesium, and then more basic.

The amorphous and para-crystalline fraction is a little more abundant than in the former soil: ~ 12 % in the A_1 horizon, 9 % in the A_3, 7 % in the B_1. Nevertheless it decreases suddenly in the B_2 and B_3 horizons, at the bottom of the soil. The molar SiO_2/Al_2O_3 ratio of the products is near 1. That shows again an aluminous allophane; although the part of allophane is greater (9 % in A_1, 6 % in B_1) and that of complexed or crypto-crystalline Al and Fe hydroxides is lesser (~ 1-2 %). In the upper part of soil (A and B_1), the < 2 μ fraction shows mainly a fluffy gel, classic for granular allophane. There are only traces of clay minerals: (10-14 Å) altered mica, smectite and (10 Å) halloysite. But at the bottom, in the brownified cinders, some sphaeroidal and tubular (10 Å) halloysite appears suddenly in a great quantity (10-20 %), while the allophane decreases (to 2-1 %).

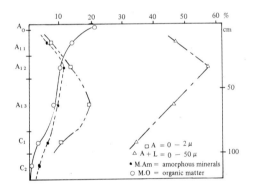

Fig. 1:   Distribution of organic matter, amorphous minerals and < 50 μ, < 2 μ fractions, in profile 12.

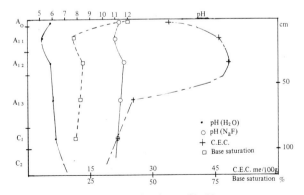

Fig. 2:   Variation of pH, CEC and base saturation in profile 12.

5. PHYSICAL AND CHEMICAL PROPERTIES

5.1.   MTE FOGLIANO (Fig. 1 and 2)

5.1.1.   Physical properties

The soil is loamy (46-56 % of $< 50\,\mu$m) and very humiferous until a depth of 80 cm. The organic matter content decreases gradually, from 14 % in A_{11}, to 7 % in A_{13}, while the clay ($<$ 2 μm) increases from 10 to 20 %. The amorphous mineral fraction increases first from 6 to 11 %, and after decreases in parallel with the organic fraction. The water retention capacity of air-dried soils is only $\sim$ 20 % at 15 bars and $\sim$ 30 % at $\frac{1}{3}$ bar. These low values are probably due to the drying effect. The structural stability is strong (Henin's index Is $<$ 1).

5.1.2.   Chemical properties

The (H_2O)pH is very acid, $\sim$ 5, in the topsoil; afterward it rises gradually to 7, at the bottom (C_2). The (NaF) pH is always $>$ 11. The (pH 8) CEC value is high, near 45 me/100 g in the A_{11} - A_{12} horizons, and decreases to 25 me/100 g in A_{13} and C_1. But the $\triangle$CEC value is very high, near 25 me/100 g in A_{13}, where the rate of $\triangle$ CEC is $\sim$ 80 %. This value and the ZPC $\sim$ 5 (in A_{13} soil without organic matter) characterize some aluminium rich allophanic products. The base-saturation is near 20 % in the whole profile. The available phosphorus is less than 10 % of total phosphorus (P-retention capacity = 95 %). The exchangeable Al^{3+} is $\sim$ 1.6 - 1.7 me/100 g in the A_{12} – A_{13} horizons.

5.2.   MTE VENERE SOIL (Fig. 3 and 4)

5.2.1.   Physical properties

The soil is loamy (44 - 54 % of $< 50\,\mu$m) and very humiferous (9 - 6 % of Org. Mat.) until a depth of 50 cm. Afterward, it becomes apparently more sandy (70 % of $> 50\,\mu$m) and less humiferous (2 - 1 % of Org. Mat.). However the $< 2\,\mu$m fraction remains near 10 - 12 % for the main part of soil, except the B_3 horizon at the bottom where the clay increases to 26 %. The amorphous mineral fraction is near 12 % and maximum in the humic horizons; afterward it decreases gradually to 3 % in parallel with the fall in humus content, in the $B_2 - B_3$ horizons, where some clay minerals (halloysite) are formed in abundance. The water-retention capacity of air-dried soil is near 20 % at 15 bars and $\sim$ 30 - 25 % at $\frac{1}{3}$ bars. The structural stability is very good (Is $<$ 0.5).

5.2.2.   Chemical properties

The (H_2O) pH is slightly acid: $\sim$ 6.3 - 6.4 in the humic horizons, rising gradually to 7 in B_2 – B_3. The (NaF) pH is $\sim$ 12-11 in the A and B_1 horizons, afterward it decreases to $\sim$ 10 in B_2 – B_3. The (pH 8) CEC value is high: $\sim$ 45 - 35 me/100 g in the humic horizons and $\sim$ 20 me/100 g in the mineral horizons. The $\triangle$ CEC value is only 10 me/100 g in the B_1 horizon, in spite of its allophane content. However the rate of $\triangle$ CEC $\sim$ 55 % and the ZPC $\sim$ 4.4

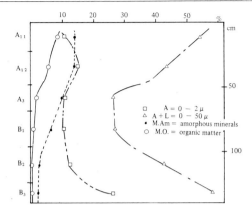

Fig. 3:   Distribution of organic matter, amorphous minerals and $< 50\,\mu$, $< 2\,\mu$ fractions in profile 13.

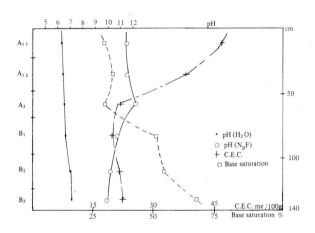

Fig. 4:   Variation of pH, CEC and base saturation in profile 13.

characterize well a moderately aluminous allophanic product. The base saturation is near 30 % in the humic horizons, and afterward rises progressively to 50 and 65 % in the B_2 and B_3 horizons. The exchangeable Al^{3+} is very low. The rate of available phosphorus is less than 15 % of the total phosphorus (P-retention capacity $= 95 - 88$ %).

6.   ORGANIC FRACTIONS AND ORGANO–MINERAL COMPLEXES

6.1.   MTE FOGLIANO SOIL (Fig. 5 and 6)

N.B. To interpret the organic fractions, see the diagram in LULLI et al., 1983, p. 46.

a)   The decrease in organic matter content is fairly slow through the whole profile. Nevertheless the humification is very high (AH + AF $\sim$ 75 - 80 % of total C). Since the A_{11} horizon, the plant residues (ML) are very few. Three organic fractions are predominating:

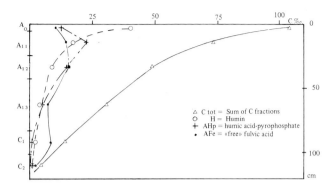

Fig. 5:   Distribution of organic fractions as C % of < 2 mm soil, in profile 12.

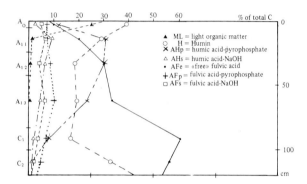

Fig. 6:   Relative evolution of organic fractions as % of total Carbon, in profile 12.

humin, "free" fulvic-acid (PO_4H_3 extract) and "pyrophosphate extractable"-humic-acid. The humin seems essentially an inherited fraction, since it decreases regulary through the profile, until it becomes stable in the C horizons, where humin is enriched by a form of "condensation". The pyrophosphate-humic acids are polycondensed humic products, that are stabilized by complexation with Al and Fe. These products increase with depth in the upper part of humic soil, where they are forming in abundance and predominating; afterward they decrease progressively. The free-fulvic acids are little condensed and unstable forms of humus, that give chelates with Al. They increase with depth till the bottom of A_{12} horizon, and afterward decrease slowly. Nevertheless in relative value, their proportion is still increasing, so that the free-fulvic acids become predominating in A_{13} and C_1 horizons, until in C_2, where they begin to be changed likely in humin by condensation.

    b)   The analysis of Al and Fe extracted by four specific reagents (dithionite, oxalate, pyrophosphate, tetraborate, after JEANROY 1983) shows the following data. The free (non-allophanic) aluminium (pyrophosphate-extractable) is abundant. It decreases slowly from 1.2 to 0.9 %, through the $A_1 - A_{12}$ horizons, and falls suddenly to 0.4 % in C_1. The whole free Al is complexed and almost 100 % chelated (in the tetraborate extract) in the whole soil profile. The free Fe is less, decreasing progressively from 0.8 to 0.4. Moreover the complexed part of iron is restricted: $\sim$ 64 % in A_{12}, 26 % in C_1. It is not truly chelated since it is not in the tetraborate extract, but it is rather as pseudo-complexed oxyhydroxides in the pyrophosphate extractable-humic acid forms.

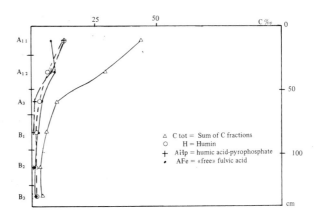

Fig. 7:   Distribution of organic fractions as C % of < 2 mm soil, in profile 13.

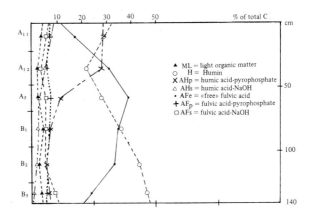

Fig. 8:   Relative evolution of organic fractions as % of total Carbon, in profile 13.

6.2. MTE VENERE SOIL (Fig. 7 and 8)

a)   The humification is very important (AF + AH $\sim$ 70 - 75 % of total C). Few plant residues remain. The organic matter decreases gradually with depth in the A_{11} – A_{12} horizons, then quickly in A_3 – B_1, and it is finally stabilized to 0.4 % of C in B_2 – B_3. The three main fractions are still humin, free-fulvic acids, pyrophosphate-humic acids. The humin decreases slowly. Nevertheless in relative proportion of total carbon, the humin increases since A_3 to B_2 and B_3 horizons, where it is predominating. As the humin decreases with depth, there is a corresponding increase in the pyrophosphate-humic acids, which derive of the humin. Indeed, these humic acids are growing and stabilized as Al-complexes in the A_{11} – A_{12} horizons, and afterward they change of form and decrease quickly. The free-fulvic acids increase clearly from A_{11} to A_{12}, and afterward decrease slowly. But, in proportion of total C, they increase even in A_3 and B_1, where they are predominating, and afterward decrease rapidly to turn in humin by condensation.

b)   The diverse forms of extractable Al and Fe by 4 specific reagents are the followings. There is a little of free non-crystallized iron, $\sim$ 0.5 - 0.4 % in the whole soil, but it is almost non-complexed and therefore at an oxy-hydroxide state. The oxalate extractable alumina is mainly engaged in the allophane structure. But there is a little of free alumina: $\sim$ 0.5 % in A_{12}, 0.3 % in A_3 and 0.2 % in B_1. This non-allophanic alumina is half and half in chelated (tetraborate-extractable) and pseudo-complexed (pyrophosphate-extractable) forms. These data differ greatly with that of Mte Fogliano soil.

7.   INTERPRETATION

7.1.   ANDIC CHARACTERISTICS

Both soils have typical andic characteristics for the most part of their profile: a loamy and very humiferous texture, a low bulk density, the lack of evident clay minerals, mostly amorphous weathering products, a water retention capacity at 15 bars $>$ 1/1 g of $<$ 2 μ fraction, (NaF) pH $>$ 10, CEC $>$ 1 me/1 g of $<$ 2 μ fraction, a rate of $\triangle$ CEC $>$ 50 %, ZPC $\sim$ 4.4-5, base saturation $<$ 50 %, Phosphorus retention rate $>$ 80 %, high content in Al-complexed humic and fulvic acids. These properties characterize the andosols which are formed under a regularly wet climate and are unsaturated in exchangeable bases. These soils remain still slightly weathered, according to the great quantity of residual primary minerals. But the colloidal fraction is largely dominated by aluminium-rich allophanic products and organo-mineral complexes. However, at the soil bottom, the disappearance of organo-mineral complexes coincides with the diminution of allophane to the benefit of halloysite formation.

7.2.   DIFFERENCES BETWEEN THE TWO ANDOSOLS

a)   The Mte Fogliano soil shows an accumulation of raw organic matter on the surface and a deep penetration of humic and fulvic acids, without the formation of B horizon. The soil is acid, rich in Al and Fe organic complexes, the importance of which equals at least that of allophane. The aluminium is predominant in the weathering products, as well as chelates as proto-imogolite. That explains the $\triangle$ CEC rate $\sim$ 80 % and the ZPC $\sim$ 5. In addition, there is a relative enrichment in free fulvic acid to the bottom of humic horizons (where C/N ratio is 16-17). This soil presents in the same time some andic and podzolic features; but the latter are not conspicuous, due to the lack of iron migration, as it seems to be the same case in cryptopodzoliques-humifères" soils (C.P.C.S. 1976). These transitional characteristics could be due to the peculiar conditions of soil formation: a temperate-perhumid climate with a cold winter, under a beech-forest, from trachytic ash (rich in silica and alumina). These peculiarities are near that of a similar andosol in the Mte Vulture (LULLI et al. 1983).

b)   The Mte Venere soil shows a very clear B horizon, but no raw organic matter accumulation. The penetration of humic and fulvic acids is limited to the upper part of soil, although few free fulvic acids move till the top of B horizon. The soil is slightly acid and contains only few chelates, but rather pseudo-complexed forms of Al and Fe by humic acids. Moreover, the allophane is predominant among the weathering products, and its shape is rather globular than fibrous. The presence of Fe oxy-hydroxide, though at a crypto-crystalline state and in traces, becomes evident in the B horizons, while halloysite begins to form. The ratio of $\triangle CEC \sim 55\%$ and ZPC ~ 4.4 show that this soil contains less free alumina than the Mte Fogliano soil. Therefore the Mte Venere andosol is more typical. This fact could be explained by a sunnier climate, under an oak-forest, and a slightly more basic parent material.

7.3. CLASSIFICATION

The characteristics of both andosols lead to their following classification in the 1: French (1972), 2: American (1975) taxonomies and 3: after ICOMAND (1983) proposals.

	Mte Fogliano andosol	Mte Venere andosol
1:	Andosol peu différencié, humique, mélanique, intergrade crypto-podzolique	Andosol différencié, désaturé non perhydraté, chromique, modal
2:	Typic Dystrandept? or andic Haplumbrept?	Typic Dystrandept
3:	Typic Melanalland	Typic Hapludand

7.4. GENESIS OF ALLOPHANE AND HALLOYSITE

The distribution of allophane and halloysite through the profile of both andosols shows that their formation is controlled by the chelating reaction of free fulvic acid and the pseudo-complexing effect of humic acid.

In the Mte Fogliano humic andosol, while Al is largely chelated, the formation of allophane is delayed. This increases to the bottom of the humic horizons. The proto-imogolite, an Al-rich form of allophane, is predominant. This form is stabilized by adsorption of humic acids. Finally, in the C_2 horizon, the disappearance of complexing agents allows the formation of halloysite.

In the Mte Venere chromic andosol, Al is poorly chelated. Therefore the allophane is abundant from the top of soil. The globular shape, a less aluminous form of allophane, is predominant. It is stabilized by humic acid in the humic horizons, and afterward decreases progressively in the top of the B horizons, while halloysite begins to appear. At the soil bottom, in B_2 and B_3, the halloysite is forming in abundance, since the penetration of complexing agents is stopped. This formation could be developed from some silica in solution and from alumina which is released after the transformation of organic complexes.

8. CONCLUSION

Two types of andosol have been observed on the volcano of Vico. The "humic and

melanic" one, in Mte Fogliano, is formed under the coldest and wettest climate. It is characterized by a deep penetration of complexing organic acid and moreover of Al-chelate, which
retards the formation of allophane and favours the formation of proto-imogolite. This soil
type is akin to the "cryptopodzoliques-humifères" soils and is classified as a Melanalland. The
"chromic and unsaturated" type, in the Mte Venere, is located under a sunnier and less wet
climate. The penetration of humic acid is limited. Moreover, the almost lack of Al-chelate
allows the formation of allophane in abundance, afterward the differentiation of a rubefied B
horizon, and finally, at the soil bottom, the abundant formation of halloysite. This typical
unsaturated and chromic andosol is classified as a Hapludand.

BIBLIOGRAPHY

C.P.C.S. (1967): Classification des sols. I.N.A., Paris.
FARMER, V.C. et al. (1978): Clays Minerals. 13 (3), 271-274.
GROUPE DE TRAVAIL ANDOSOLS (CPCS) (1972): Cahiers ORSTOM, sér. Pédologie, X (3),
        302-303.
ICOMAND (1983): Circular letter n. 5, New Zealand Soil Bureau, Lower Hutt.
JEANROY, E. (1983): Thèse science naturelle université Nancy I.
LULLI, L. et al. (1983): Cahiers ORTSOM, sér. Pédologie, XX (1), 27-61.
U.S.D.A. (1975): Soil Taxonomy. Soil Conservation Service, Washington.
VIOLANTE, P. & TAIT, J.M. (1979): Clay Minerals. 14 (2), 155-158.

Addresses of authors:
P. Quantin, B. Dabin, A. Bouleau, ORSTOM, 70 route d'Aulnay,
93140 Bondy, France
L. Lulli, D. Bidini, Istituto Sperimentale per lo Studio e la Difesa del Suelo, Piazza d'Azeglio 30
50121 Firenze, Italy

E. Fernandez Caldas & Dan H. Yaalon (Eds):
VOLCANIC SOILS
CATENA SUPPLEMENT 7, Braunschweig 1985

GENESIS OF SOILS
AFFECTED BY DISCRETE VOLCANIC ASH INCLUSIONS,
ALBERTA, CANADA

A. **Limbird**, Calgary

SUMMARY

Soil profiles which include four discrete ash deposits were examined in the Front Range west of Calgary, Alberta. The ashes have been identified as Bridge River (about 2600 years old), St. Helens Y (about 3400 years old), Mazama (about 6600 years old), and Glacier Peak (about 12000 years old). Three of the four ash deposits occupy the positions of, and function in the manner of Ae (E) horizons. Associated with each of these three ashes are thin, distinct and strongly structured Bt horizons. The second ash from the surface (St. Helens Y) and the lowest ash (Glacier Peak) have thin Ah horizons associated with them as well, while the surface ash (Bridge River) has an associated litter layer (LFH horizon). The Mazama ash is the only one which does not have an associated soil profile. The area has undergone at least three periods of landscape stability during which soil formation has occurred as indicated by the surface profile and two buried profiles. The dated ashes demonstrate a rate of soil development which has been enhanced markedly by the ashes.

1. INTRODUCTION, MATERIALS, METHODS

Soil profiles which include four discrete volcanic ash deposits were examined at sites in the Kananaskis Valley in the Front Range of the Rocky Mountains (Figure 1). The site locations are about 85 kilometers (50 miles) west of Calgary, Alberta, Canada. The area has an average annual temperature of 5°C and average annual precipitation of about 600 millimeters. The pedon sites are situated at about 1300 meters elevation along the leeward edge of a large alluvial fan where it comes in contact with a much smaller fan from an adjacent creek. The sites have a very gentle slope (< 3°) to the west and are on the northeast side of the large fan and the southwest side of the small fan. The soils of the area are mapped and classified as Orthic Eutric Brunisols (Orthic Eutrochrepts).

The sites in this study are very unusual because of the preservation of four ashes at such great distances from their sources (Figure 1). The ashes have been identified as: Bridge River ash which is between 2440 + 140 years old (NASMITH et al. 1967) and 2670 + 140 years old (WESTGATE & DREIMANIS 1967). Its origin is at Meager Mountain in the southern Coast Range of British Columbia about 700 kilometers from the Kananaskis Valley. St. Helens Y ash which is older than 2980 + 250 years old (CRANDELL et al. 1962) and younger than 3550 + 65 years old (WESTGATE et al. 1969). Its origin is at Mount St. Helens in the state of Washington about 800 kilometers from the study area. Mazama ash which is between 6290 + 250 years old and 6630 + 200 years old (POWERS & WILCOX 1964, WESTGATE et al. 1969). Its origin is at Crater Lake in the state of Oregon (Mount Mazama) about 1100 kilometers from the study area. Glacier Peak ash which is 11250 years old (MERHINGER 1977) to 12000 years old (FRYXELL 1965). Its origin is at Glacier Peak in the state of Washing-

ISSN 0722–0723 / ISBN 3–923381–06–9

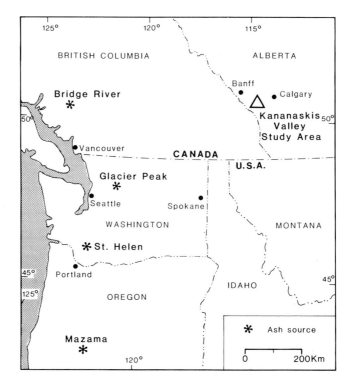

Fig. 1:   Location of Study Area

ton about 600 kilometers from the study area.

The magnetite grains present within an horizon is the chief indicator of volcanic origin. Analysis of individual glass-encased magnetite grains provides the basis for identification of the ash fall and source from which the ash layer was derived. The results are compared with values for site specific flows (Table 1) identified based on examination of Al203, TiO2, and MgO concentrations (BREWSTER & BARNETT 1979, WESTGATE et al. 1970, WEST-GATE & FULTON 1975, WESTGATE 1977).

The soil profiles have been studied in some detail because three of the four ash deposits occupy the position of, and function in the manner of, Ae (E) horizons with associated thin,

Tab. 1:   MAGNETITE RATIOS OF VOLCANIC ASHES AT STUDY SITES COMPARED TO KNOWN ASH SAMPLES

	$Al_2O_3:TiO_2$	$MgO:TiO_2$
Known Bridge River Ash	1.87:7.56	1.99:7.56
Study Site Bridge River Ash	1.98:7.45	1.85:7.45
Known St. Helen Y Ash	2.61:5.85	1.18:5.85
Study Site St. Helen Y Ash	2.50:6.00	1.20:6.00
Known Mazama Ash	2.09:8.66	2.40:8.66
Study Site Mazama Ash	2.20:9.10	2.45:9.10

Note: Glacier Peak ash identified by stratigraphic position below Mazama Ash.

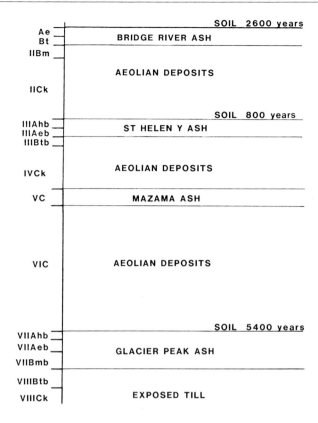

Fig. 2: Diagram of Ashes and Soil Horizon Development

distinct, and strongly structured Bt horizons. The second ash from the surface (St. Helens Y) and the lowest ash (Glacier Peak) have thin Ah horizons associated with them as well, while the surface ash (Bridge River) has an associated litter layer (LFH horizon). These soil profiles indicate that the area has undergone at least three periods of landscape stability during which soil formation occurred as shown by the surface profile and the two buried profiles. The time framework of these dated ashes helps to express the maximum duration for each period of pedogenesis (Figure 2). The analysis data presented suggest that the ashes have functioned to speed up the processes of soil development in the Kananaskis Valley when compared to soils unaffected by the ash falls.

Soil samples from each horizon were analyzed for texture using the hydrometer method, pH levels using a 2:1 water:soil saturation, conductivity using a 2:1 water:soil saturation, organic matter using potassium dichromate and sulfuric acid with spectrophotometer, total mineral composition and clay minerals by xray diffraction, cation exchange capacity using 1N ammonium acetate and atomic absorption, and iron and aluminum using pyrophosphate extraction and atomic absorption. The analyses followed the procedures outlined in McKEAGUE (1978).

2. RESULTS AND DISCUSSION

The soil profiles of the study sites have unusually well preserved clear boundaries which indicate a paucity of biotic mixing of horizons. Various species of ants and earthworms are found in soils in the Kananaskis Valley. However, none have been observed in the study site soils. The absence of biota in these profiles is somewhat abnormal and cannot be explained in the context of this study. Recent windthrow of trees is fairly common which may help to explain the less abrupt boundaries of the surface soil profiles and to support the interpretation of more recent forest cover compared to the buried profiles.

In the profile description the volcanic ash inclusions are identified as the Ae, the III Aeb, the V C, and the VII Aeb horizons (Table 2). Each of the subsurface ash layers has an abrupt, smooth boundary to the underlying material indicating that the ash fell on an unobstructed surface. There is no evidence that these ashes fell on already existing soil horizons unless the profiles were truncated by erosion. Erosion does not seem very likely because of the overall sequence of depositional materials at the sites. Thus, soil horizon development began in the cases of the Glacier Peak and St. Helens Y ashes after ash fall. The gradual, wavy lower boundary of the Bridge River ash suggests the ash may have fallen on an already existing soil surface. In fact, the author has found locations in the Foothills region near Calgary where there

Tab. 2:   PROFILE DESCRIPTION OF SOIL WITH FOUR VOLCANIC ASH INCLUSIONS

LFH	4 to 0 cm. Burned and partly decomposed forest litter.
Ae	0 to 5 cm pinkish-gray (7.5 YR 7/2) silt loam; moderate, thin platy structure; very friable; few fine roots; gradual, wavy boundary; Bridge River ash inclusions.
Bt	5 to 10 cm dark brown (7.5 YR 4/4) silty clay; strong, medium subangular blocky structure; firm to very firm; moderate fine and medium roots between peds; clear, smooth boundary.
II Bmk	10 to 17 cm brown (10 YR 5/3) loam; moderate, fine crumb to weak fine platy structure; friable; few fine and medium roots; abrupt smooth boundary; weakly calcareous.
II Ck	17 to 55 cm pale brown (10 YR 6/3) loam; moderate, fine granular structure; friable; few roots; abrupt smooth boundary; strongly calcareous.
III Ahb	55 to 58 cm very dark brown (10 YR 3/2) silt loam; moderate fine granular structure; very friable; few roots; clear, smooth boundary.
III Aeb	58 to 62 cm very pale brown (10 YR 7/3) silt loam; moderate, fine platy structure; very friable; no roots; abrupt, smooth boundary: St. Helens Y ash inclusion.
III Btb	62 to 67 cm brown (7.5 YR 5/4) silty clay; strong, fine angular blocky structure; very firm, few fine roots; clear, smooth boundary.
IV Ck	67 to 98 cm gray (10 YR 6/1) silt loam; massive, breaking to strong, fine angular blocky structure; extremely firm; no roots; abrupt, smooth boundary; strongly calcareous.
V C	98 to 102 cm white (10 YR 8/1) silt; packed but loose lacking structures; no roots; abrupt, smooth boundary; Mazama ash.
VI C	102 to 170 cm light gray (10 YR 7/2) loam to light brownish gray (10 YR 6/2) sandy loam in stratified layers; very weak, fine granular structure; friable; no roots; clear, smooth boundary; moderately calcareous.
VII Ahb	170 to 172 cm grayish brown (10 YR 5/2) silt loam; very weak, fine granular structure; friable; no roots; clear, smooth boundary.
VII Aeb	172 to 180 cm light gray (10 YR 7/2) silt loam; weak, very fine granular structure; very friable; no roots; abrupt, smooth boundary; Glacier Peak ash inclusion.
VII Bmb	180 to 190 cm brownish yellow (10 YR 6/6) silt loam; weak, very fine granular structure; friable; no roots; clear, smooth boundary.
VIII Btb	190 to 202 cm yellowish brown (10 YR 5/6) silty clay; moderate, medium subangular blocky structure; firm to very firm; gradual boundary.
VIII Ck	202 cm+ brownish yellow (10 YR 6/6) silty clay loam to clay loam, strong, coarse subangular blocky structure; very firm, dense glacial till; moderately calcareous.

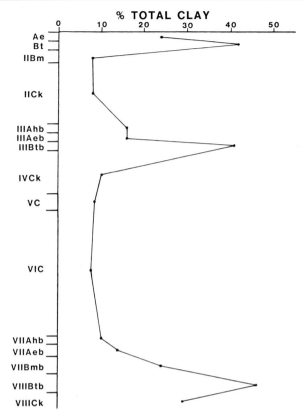

Fig. 3:   Total Clay Percentages by Horizons

Tab. 3:   PROFILE PROPERTIES OF SOIL WITH FOUR VOLCANIC ASH INCLUSIONS

Horizon/Property	%Sand	%Silt	%Clay	%Organic Matter	CEC	2:1 H₂O:Soil pH	(mmhos/cm) Saturation Conductivity	pyro-phosphate Extractable %Fe	%Al
Ae*	20	56	24	0.6	9.1	6.25	.045	.08	.13
Bt	8	50	42	2.7	23.3	6.85	.128	.27	1.10
II Bmk	46	46	8	0.3	5.0	7.70	.260	.14	.34
II Ck	49	43	8	0.3	5.0	7.81	.550	.06	.10
III Ahb	8	76	16	4.1	17.2	7.65	.248	.05	.06
III Aeb*	8	76	16	0.6	6.1	7.46	.115	.06	.11
III Btb	15	44	41	0.5	15.6	7.61	.350	.23	.84
IV Ck	31	59	10	0.2	5.8	8.00	.106	.04	.06
V C*	10	81.5	8.5	0.0	3.2	7.95	.167	.36	.60
VI C	58	34.5	7.5	0.3	5.4	7.96	.130	.04	.06
VII Ahb	16	74	10	3.2	12.4	8.08	.090	.12	.20
VII Aeb*	2	84	14	0.3	5.3	7.52	.088	.07	.09
VII Bmb	21	55	24	0.5	9.1	7.74	.070	.26	.68
VIII Btb	7	47	46	0.8	17.5	6.87	.119	.48	.89
VIII Ck	20	51	29	0.2	11.0	7.76	.062	.08	.11

*   Volcanic Ash Inclusion Horizons.

are two Ae horizons at the surface, the upper one of ash and the lower one ash free, and abutting one another. There is no soil profile development associated with the Mazama ash which indicates either a period of instability for soil formation or a more continuous deposition of further aeolian material following the ash fall.

The volcanic ash horizons in the example profile have high silt contents (Table 3). Other researchers (KING & BREWSTER 1976, OTTERSBEG & NIELSEN 1977, WESTGATE & DREIMANIS 1967) have described the silt content of ash as greater than 60%. In keeping with this value, the III Ahb (76 %), the III Aeb (76 %), the V C (81.5 %), the VII Ahb (74 %), and VII Aeb (84 %) horizons are recognized as part of the volcanic ash inclusions. The Ah horizons are ashes which have been enhanced by the incorporation of organic matter. The surface ash Ae horizon contains less silt which is the result of mixing and resorting of the ash with other wind blown materials or more intense weathering of silt to clays under more recent climatic conditions. The aeolian material is found in small and scattered deposits in the Front Range and Foothills areas except along the major river corridors traversing these areas. In the Kananaskis Valley the aeolian material is considered to be of local origin from the floodplain of the Kananaskis River. The high calcite and dolomite content of the aeolian material is similar to the local alluvial, colluvial, and till parent materials of accompanying soils of the valley.

Strongly developed B horizons are associated with the three ash layers that have soil development (Table 2). The Bt horizon with 42% clay and strong subangular blocky structure and very firm consistence stands out in sharp contrast to its adjacent horizons despite being thin (Figure 3). The III Btb horizon with 41% clay is similar in many ways to the Bt horizon, especially the strong structure and very firm consistence. The VIII Btb horizon with 46% clay also has strong structural development and very firm consistence, but is influenced somewhat more by the underlying glacial till material. However, in each case the development of the B horizon seems to have been enhanced by the porous nature of the volcanic ash material above, which has allowed for finer materials to move in suspension to the B horizon where subsequently weathered clays would collect and form in situ from the ash minerals.

The pH levels of the profile are quite high (Table 3) and can be explained by the calcareous nature of the windblown materials in the Kananaskis Valley. Much of the surrounding till, colluvium, and alluvium has originated from limestone bedrock and has not been highly leached because of the moderately dry climatic conditions. The Kananaskis Valley is on the leeward side of the mountains from the prevailing westerly winds which bring most of the precipitation from the Pacific Ocean. The pH of the surface Ae horizon is related to a more recent increase in leaching intensity. Otherwise, the overall trend suggests a slight pH depression associated with the volcanic ash inclusions; the III Aeb (7.46) and the VII Aeb (7.52) do have pH values lower than the adjacent horizons. However, the presence of highly calcareous aeolian and fluvial materials in the area of the profile sites suggests a reintroduction of calcareous materials into the horizons with each new parent material, except for the ashes. In fact, field testing with 2N hydrochloric acid indicated that only the II Ck, IV Ck, VI C, and the VIII Ck horizons had moderate to strong reactions. However, except for the VIII Ck horizon, these materials are interspersed among the developed horizons in such a way as to introduce more carbonates with each depositional event and thus mask previous leaching and profile weathering.

Conductivity values indicate a general decrease of ions in solution with increasing depth in the profile (Table 3). The extreme value in the II Ck horizon suggests a calcareous aeolian origin for this deposit which may have been enhanced by subsequent leaching from above. There is an increase of ions in solution in the Bt horizons relative to the overlying Ae ashes because of relatively rapid leaching through the ash and the accumulation of ions in the presence of clays in the Bt horizons.

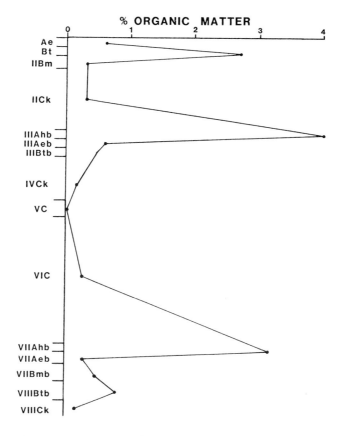

Fig. 4:   Organic Matter Content by Horizons

Tab. 4:   IMPORTANT MINERALS OF HORIZONS OF PROFILE WITH FOUR VOLCANIC ASH
INCLUSIONS

Horizon	Bulk Sample Minerals	Clay Fraction Minerals
Ae	Quartz, Feldspar, **Chlorite**	Illite, Kaolinite, Chlorite, some **Allophane**
Bt	Quartz, Dolomite, Feldspar	Illite, Kaolinite, Chlorite, some **Allophane**
II Bm	Dolomite, Quartz, Felspar	Illite, Kaolinite, Chlorite
III Ahb	Quartz, Dolomite, Feldspar	Illite, Kaolinite, Chlorite
III Aeb	Quartz, **Calcite**, Dolomite	Illite, Kaolinite, Chlorite, some **Allophane**
III Btb	Dolomite, Quartz, Feldspar	Illite, Kaolinite, Chlorite, some **Allophane**
VII Ahb	Dolomite, Quartz, **Calcite**	Illite, Kaolinite, **Vermiculite**
VII Aeb	Dolomite, **Calcite**, Quartz	Illite, Kaolinite, **Vermiculite**, some **Allophane**
VII Bmb	Quartz, K Feldspar, Na Feldspar	N.A.
VIII Btb	Quartz, Feldspar, **Mica**	N.A.

Note: ..... minerals relate to prominent mineral shifts related to ashes or to underlying glacial till.
N.A. = Not analyzed.

Organic matter was analyzed for the horizons of the profile after pretreatment with hydrochloric acid to remove carbonates. Only the Bt, the III Ahb, and the VII Ahb horizons contain appreciable amounts of organic matter (Figure 4). The Bt horizon organic matter originated from the surface LFH horizon and has passed through the Ae horizon by relatively intense leaching. The degree of mobility of the organic matter may relate to the recent climatic past with a tendency to podzolization and/or to the present vegetation of *Pinus contorta* (Lodgepole Pine) and *Populus tremuloides* (Quaking Aspen). The isolation of organic matter in the two Ahb horizons confirms them to be former surface horizons. The incorporation of organic matter and the clear smooth boundary to the underlying ash horizons confirms them to be Ah rather than LFH horizons. The thinness of these two horizons (III Ahb and VII Ahb) suggests either some partial truncation which is ruled out in the depositional environment or a relatively short term grass community which would be the consequence of fires or somewhat drier climatic conditions than at present. The lack of organic matter in other horizons suggests a relatively short time span for a vegetative cover. Otherwise, at least some organic matter would have been translocated to the underlying B horizons.

Only the A and B horizons were analyzed for mineral composition. Bulk samples were run using ions, 40000 Kv, and 20000 amps. Then the clay fraction was separated from each sample, saturated with water, and then with calcium chloride. Untreated clays were run on the xray and then samples were treated with ethylene glycol, heated at 500 C for one hour and rerun.

Quartz and feldspars are abundant in all the bulk samples (Table 4) for all the horizons except for the St. Helens Y ash inclusion (III Aeb) and the Glacier Peak ash inclusion (VII Aeb) where calcite replaces feldspar in importance. Dolomite is abundant in all the bulk samples for all horizons, except for the Bridge River ash inclusion (Ae) where chlorite is more important and the VIII Btb horizon, which is influenced by the relatively unweathered underlying glacial till, where mica is more important. The ash inclusions are recognized by the shift in the most important minerals compared to the other horizons in the profile. In addition, magnetite which is recognized as being present in volcanic ash, is distinguished in the ash inclusion horizons by a 2.53A peak on the xray diffractograms.

The use of calcium chloride in the clay fraction analysis was especially helpful in distinguishing the twin peaks of chlorite and kaolin at 7.1A. Glycolating the slides resulted in expansion of illite peaks and made the identification of montmorillonite, kaolinite, and mixed layer clays possible. The slides which had been prepared in a water medium were heated to 500 C for one hour following glycolation. Heating collapsed the kaolin peak at 7.1A but left a peak if chlorite was present. If the peak at 14A remained after heating, chlorite was present, but if it disappeared vermiculute was present.

The presence of chlorite in the upper horizons is in agreement with other researchers in the Canadian Rocky Mountains (KING & BREWSTER 1976) and seems to originate with the aeolian surface deposits of this region. Chlorite decreases substantially in the lower horizons while vermiculite increases which suggest an increased influence of the proglacial deposits, especially the underlying till.

Illite, kaolinite, and montmorillonite are present in all horizons (Table 4) and in most cases these secondary minerals are well developed. The weathering to produce these minerals has been substantial, implying that the horizons comprising the profiles are aged pedologically faster than their chronological ages would suggest. Mixed layer clays are prominant also throughout the profiles. These interstratified clays are either chlorite-vermiculite or illite intergrades. These intergrade clays also indicate the well developed character of the profiles. Allophane is present only in small amounts and only in the ash horizons (Ae, III Aeb, VII Aeb)

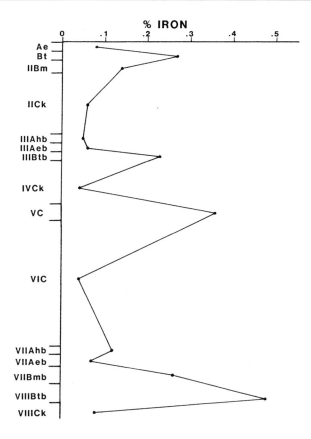

Fig. 5: Iron Content by Horizons

and ash associated horizons (Bt, III Btb).

The cation exchange capacity of the horizons follows a pattern of close relationship of cation exchange values with clay and especially organic matter content of the individual horizons. The Bt horizon and the III Ahb horizon have high CEC values due to the combination of high clay and organic matter; and the III Btb and VII Btb horizons have high CEC values due to high clay content. The volcanic ash inclusion horizons have relatively low CEC values because of the virtual lack of organic matter and low clay contents. Several trends are apparent in the cations present. Calcium dominates the exchange complex, but is particularly concentrated in the C horizons (except for the VC, Mazama ash). The volcanic ash inclusion horizons have reduced concentrations of cations which suggests translocation from these horizons. However, the B horizons do not show as marked an increase in cations as might be expected based on clay content. Other researchers (GOLDIN 1983, KING & BREWSTER 1976, OTTERSBERG & NIELSEN 1977) have shown that CEC values are a function of organic matter and not of clay distribution. In addition, the relatively intense weathering as shown by the clay mineral assemblage suggests a possible reduction in CEC in proportion to clay content as the clays have weathered to contain at least some kaolinite which has a low CEC. Available potassium values are lowest in the B horizons which also suggests that the potassium may be

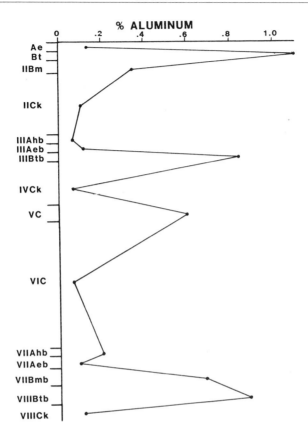

Fig. 6:   Aluminum Content by Horizons

usurped as part of the clay structures in the secondary mineral formation process.

The analysis values of iron and aluminum indicate that iron (Figure 5), and especially aluminum (Figure 6), are important elements associated with the volcanic ash inclusions. Weathering and eluviation has been strong enough to translocate the iron and aluminum into the underlying B horizons. The aluminum that concentrated in the Bt, the III Bt, and the VII Btb horizons may have been instrumental in the development of secondary clays in situ, as suggested by other researchers (KING & BREWSTER 1976, SINGER & UGOLINI 1974) rather than through clay translocations from above. The Ae horizon ash inclusions have facilitated the movement of clay structural components to the B horizons, especially aluminum. Water percolates easily to hydrolyze the ash and create the amorphous materials which are translocated and then converted to clays in the Bt horizons. The rate of clay development is much faster than would occur without the ashes being present. The soil associated with the Glacier Peak ash is classified as a Brunisolic Gray Luvisol (Eutroboralf); the soil associated with the St. Helens Y ash is classified as an Orthic Gray Luvisol (Eutroboralf); and the soil associated with the Bridge River ash is classified as a Brunisolic Gray Luvisol (Eutroboralf) (STRONG & LIMBIRD 1981). In contrast, the surrounding soils which are unaffected by ash inclusions are Orthic Eutric Brunisols (Eutrochrepts) and do not contain Bt horizons.

3. CONCLUSIONS

Under the environmental conditions of the Kananaskis Valley the volcanic ash hydrolyzes rapidly aided by the high porosity and permeability of the ash and accompanying aeolian material. The leaching regime is enhanced by moisture from snow melt and has the potential for producing considerable amounts of amorphous secondary weathering products fairly rapidly. However, considering the short time available for the development of each sub-profile, the clay mineral assemblage is well advanced indicating a more rapid than usual rate of pedogenesis, especially Bt horizon development, as a result of the volcanic ash inclusion. Using generalized dates, the soil profile of the Glacier Peak ash had to develop in less than 5400 years, the soil profile of the St. Helens Y ash had to develop in less than 800 years, and the soil profile of the Bridge River ash had to develop in less than 2600 years (Figure 2). The typical soils of the surrounding Kananaskis Valley have Ae horizons and only moderately developed Bm (cambic) horizons. These Brunisols (Ochrepts) have been exposed to pedogenesis for up to 12000 years, based on the age of Glacier Peak ash in contact with unweathered glacial till, and yet are much less intensively developed.

BIBLIOGRAPHY

BREWSTER, G.R. & BARNETT, R.L. (1979): Magnetites from a New Unidentified Tephra Source, Banff National Park, Alberta. Canadian Journal of Earth Science **16**, 1294-1297.

CRANDELL, D.R., MULLINEAUX, D.R., MILLER, R.D. & RUBIN, M. (1962): Pyroclastic Deposits of Recent Age at Mount Ranier, Washington. United States Geological Survey Professional Paper **450-D**, 64-68.

FRYXELL, R. (1965): Mazama and Glacier Peak Volcanic Ash Layers: Relative Ages. Science **174**, 1288-1290.

GOLDIN, A. (1983): Comparison of Some Podzols and Spodosols in the Fraser Lowland, British Columbia and Adjacent Washington State. Canadian Journal of Soil Science **63**, 579-591.

KING, R.H. & BREWSTER, G.R. (1976): Characteristics and Genesis of Some Subalpine Podzols (Spodosols), Banff National Park, Alberta. Arctic and Alpine research **8**, 91-104.

McKEAGUE, J.A. (1978): Manual on Soil Sampling and Methods of Analysis. Second Edition, Canadian Society of Soil Science, Ottawa, Ontario. 212 pages.

MEHRINGER, P.J., BLINMAN, E. & PETERSEN, K.L. (1977): Pollen Influx and Volcanic Ash. Science **198**, 257-261.

NASMITH, H., MATHEWS, W.H. & ROUSE, G.E. (1967): Bridge River Ash and Some Other Recent Ash Beds in British Columbia. Canadian Journal of Earth Science **4**, 163-170.

OTTERSBERG, R.J. & NIELSEN, G.A. (1977): Recognition of Volcanic Ash Influenced Soils by Soil Scientists in Western Montana and Parts of Idaho. Soil Survey Horizons **18**, 8-13.

POWERS, H.A. & WILCOX, R.E. (1964): Volcanic Ash from Mount Mazama and from Glacier Peak. Science **144**, 1334-1336.

SINGER, M. & UGOLINI, F.C. (1974): Genetic History of Two Well-Drained Subalpine Soils Formed on Complex Parent Materials. Canadian Journal of Soil Science **54**, 475-489.

STRONG, W.L. & LIMBIRD, A. (1981): A Key for Classifying Soils to the Subgroup Level in the Canadian System of Soil Classification. Canadian Journal of Soil Science **61**, 285-294.

WESTGATE, J.A. (1977): Identification and Significance of Late Holocene Tephra from Otter Creek South, British Columbia and Localities in West-Central Alberta. Canadian Journal of Earth Science **14**, 2593-2600.

WESTGATE, J.A. & DREIMANIS, A. (1967): Volcanic Ash Layers of Recent Age in Banff National Park, Albeta, Canada. Canadian Journal of Earth Science **4**, 155-161.

WESTGATE, J.A. & FULTON, R.J. (1975): Tephrostratigraphy of Olympia Integlacial Sediments in South Central British Columbia, Canada. Canadian Journal of Earth Science **12**, 489-502.

WESTGATE, J.A., SMITH, D.G.W. & NICHOLS, H. (1969): Late Quaternary Pyroclastic Layers in the Edmonton Area, Alberta. In: S. Pawluk (editor). Pedology and Quaternary Research. Symposium of the National Research Council of Canada, 179-186.

WESTGATE, J.A., SMITH, D.G.W. & TOMLINSON, M. (1970): Late Quaternary Tephra Layers in Southwestern Canada. In: Early Man and Environments in Northwest North America, Proceedings of Second Annual Paleo-Environment Workshop of the University of Calgary Archeology Association. Students Press, Calgary, 13-34.

Address of author:
Arthur Limbird, Department of Geography, University of Calgary
Calgary, Alberta, Canada

E. Fernandez Caldas & Dan H. Yaalon (Eds):
VOLCANIC SOILS
CATENA SUPPLEMENT 7, Braunschweig 1985

POLYGENESIS ON DEEPLY WEATHERED PLIOCENE BASALT, GOMERA (CANARY ISLANDS): FROM FERRALLITIZATION TO SALINIZATION

M.L. **Tejedor Salguero**, C. **Jimenez Mendoza**, A. **Rodriguez Rodriguez** and
E. **Fernandez Caldas**,
La Laguna

SUMMARY

This paper report a study of several original pedological formations, consisting of ferrallitic soils associated with ironstone, that present very deep weathering and an extraordinary degree of development associated with a more recent process of salinization.

These soils are to be found on several plateaus in the northern area of the island of La Gomera, unaffected by volcanic eruptions in the recent Quaternary and where, consequently, the soils always present a high degree of differentiation and evolution.

These soils present an A_1 horizon, although this has frequently disappeared through erosion, the A_2 horizon appearing at the surface. The A_2 horizon is an eluvial horizon, impoverished, originated by the illuviation and weathering of clays. At the present time it is salinized and sodified, with a structure that tends to be columnar and a compaction characteristics of horizons with a relatively high sodium content in the exchange complex.

Underneath the A_2 horizon, very clayey Btg horizons are generally to be found, originated by processes of illuviation in hydromorphic conditions. All the characteristics of these Btg horizons provide evidence for the existence of processes of movilization and reorganization of sesquioxides of iron and fine material leading to the formation of ironstone (goethitic) in those regions of the landscape more favourable to deposition and the crystallization of the iron sesquioxides.

All the characteristics of these soils allowed us to establish the sequence of processes (see Fig. 5).

RESUMEN: POLIGENESIS SOBRE BASALTOS PLIOCENICOS MUY ALTERADOS EN GOMERA (ISLAS CANARIAS): SUELO FERRALITICO SALINIZADO

En este artículo se realiza el estudio de una formación edáfica de gran originalidad, constituida por suelos ferralíticos con coraza ferruginosa, que presentan una alteración muy profunda y un extraordinario grado de desarrollo, al que se ha superpuesto un proceso más reciente de salinización.

Estos suelos se encuentran situados en varias plataformas de la región norte de la isla de La Gomera, isla que no se ha visto afectada por erupciones volcánicas en el Cuaternario reciente y donde, los suelos presentan siempre un alto grado de diferenciación y evolución.

Estos suelos presentan un horizonte A_1, aunque frecuentemente haya desaparecido por erosión y aparezca en superficie el horizonte A_2. El horizonte A_2 es un horizonte eluvial, empobrecido, originado por iluviación y degradación de arcilla. Actualmente se encuentra salinizado y sodificado, presentando una estructura con tendencia columnar y una compactación característica de los horizontes con un contenido relativamente alto de sodio en el complejo de cambio.

Por debajo del horizonte A_2 es general, la presencia de horizontes de tipo Btg muy arcillosos, originados por procesos de iluviación en condiciones hidromórficas. Todas las características de estos horizontes Btg demuestran la existencia de procesos de movilización

ISSN 0722–0723 / ISBN 3–923381–06–9

y reorganización de sesquióxidos de hierro y de material fino que lleva a la formación de corazas ferruginosas (goetíticas) en aquellas zonas de paisaje más favorables a la precipitación y cristalización de los sesquióxidos de hierro.

En definitiva, todas las características de estos suelos, nos permiten establecer que en la génesis de los mismos, ha tenido lugar la secuencia de procesos (Fig. 5).

1. INTRODUCTION

La Gomera is the oldest island in the Canary Archipelago and no volcanic eruptions have taken place there in the last million years. The soils of this island invariably present a

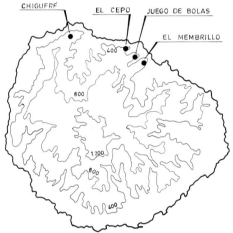

Fig. 1:   Map of Gomera.

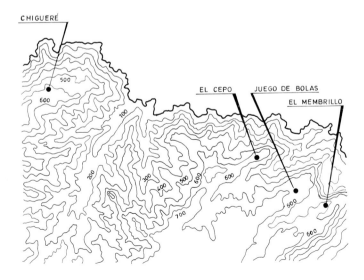

Fig. 2:   Location map of sample sites.

high degree of evolution and differentiation, and soils corresponding to recent pedological formations, such as andisols, are no observed.

On the other hand, the deep erosion that characterizes the island of La Gomera has given it a very rugged relief and a physiography peculiar to itself, with two basic features: the deep ravines that cut through the island in a radial pattern converging at the central plain of El Cedro, and the plateaus and small plains between the ravines, where the most developed and evolved soils are to be found (Fig. 1).

Pedological formation of great originality, composed of ferrallitic soils with ironstone presenting an extraordinary degree of development and a very deep weathering associated with a process of plinthization, have been found in several plateaus situated in the northern region of the island between 500 and 600 m above sea-level (Fig. 2).

1.1. ECOLOGICAL CHARACTERISTICS

The geological basement of the plateaus where the soil formation studied here are to be found is made up of the so-called "horizontal basalts" (BRAVO 1964) or "subrecent basalts" (CENDRERO 1971).

As will be shown later, all these soils were originated directly by the progressive weathering of the afore-mentioned materials.

This geological formation, consists of very thick lava flows with a clearly horizontal disposition, hence the term "horizontal basalts".

The flows are generally made up of aphyric and sometimes slightly porphyritic basalts. The mineralogic composition shows a clear predominance of plagioclase with large pyroxene, and less frequently, olivine phenocrysts. The chemical composition of these horizontal basalts, according to BRAVO (1964), is as follows:

$$SiO_2 = 42,9; \; Al_2O_3 = 14,8; \; Fe_2O_3 = 5; \; FeO = 7,6; \; MnO = 0,2;$$
$$MgO = 7,6; \; CaO = 10,1; \; Na_2O = 3,2; \; K_2O = 1,3; \; TiO_2 = 3,8.$$

The approximate age of these subrecent basalts was calculated by CENDRERO (1971) as 5 million years which places the emission of these materials in the Pliocene.

The present climatic conditions of this area, are difficult to establish due to the lack of continuous reliable climatological records. The nearest pluviometric recording station not too far away from the sampling sites, in a similar climatic environment (Meriga), gives a mean annual rainfall of 700 mm. The Thornthwaite diagram of this recording station is given in Fig. 3.

The prevailing winds blow from the north-east through out the year with varying intensity. In stormy days the spray from the sea can be very intense.

There is evidence in the Canary Islands of a more humid climatic regimen in the past. Climatic changes that affected the Macaronesian region during the Quaternary brought about a decrease in the rainfall rates and led to the milder present-day climate with a certain xeric trent.

The vegetation presently covering the plateaus consists of a very degraded and disperse fayal-brezal, poor in species. In most cases, the soils are denuded, with only a few isolated specimens of heather (Erica), eucalyptus (Eucaliptus), "vinagrera" (Rumex lunaria), "barrillas"(Mesembryanthemum), a halophytic plant that indicates salinization of the soil.

The near absence of vegetation has favoured the erosion of these soils which, in the

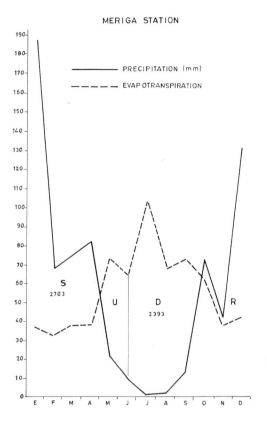

Fig. 3:   Thorntwaite diagram.

majority of the areas, are lacking in A_1 horizon, the more compact and massive A_2 horizon appearing at the surface.

The topography of these plateaus characterized by an almost flat surface sloping gently towards the North, is a consequence of the stratigraphic disposition of the geological materials. However, within each plateau the meso-relief shows great contrasts, with deep gulleys caused by the advance of erosion (Photo 1).

1.2.   MATERIAL AND METHOD

Four representative profiles of the soils existing in the northern plateaus of the island of La Gomera were selected for this study.

The pH was measured in aqueous suspension (1/2,5) and in KCl (N) (1/2,5).

Exchangeable cations and exchange capacity at pH = 7 were analysed following the method by BOWER et al. (1952).

The organic matter content was determined by the WALKEY–BLACK method (1933). The KONONOVA method (1968) was adapted for the fractioning of same.

The nitrogen, mineralized by sulphuric attack, was analysed by the Kjedahl method.

The acidity was carried out by percolation with KCl(N); the sum of $Al^{+++} + H^{+}$ was titrated with NaOH N/100 and the Al^{+++} measured on an atomic absorption spectrophotometer by Perkin Elmer model 603.

The mechanical analysis was performed following the classic robinson pipette method. The elementary analysis was carried out by HF attack and fusion with CO_3Na_2.

Conductivity in the saturated paste obtained by the RICHARDs methods (1954) was measured on a Meter LBR WTW conductimeter. The results are expressed in mmhos/cm at 25°C. The soluble cations was measured by atomic absorption, the chlorides by the MOHR method and sulphates by the BRANSON & SHARPLESS method (1957) (Ministry of Agriculture).

The determination of humidity at different values of pF were obtained by suction on porous plates, model by Soil Moisture Equipment Corporation.

The study of the primary minerals was performed according to the technique by PEREZ MATEOS (1965) and the modifications introduced by ALEIXANDRE & PINILLA (1968).

The mineralogical study by X-ray diffraction was carried out on a Philips PW 1720 diffractometer in randomly oriented and oriented samples.

2. MORPHOLOGICAL CHARACTERISTICS OF THE SOILS

The soils studied are only observed in four plateaus situated at an approximate altitude of 600 m in the northern flank of the island and show very similar ecological features.

These profiles show an intense ferrallitic-type weathering of the horizontal basalts. The weathering of these materials can sometimes be observed in the gulleys. It is interesting to note that the weathering has led to the formation of very spectacular polygonal structures (Photo 2).

The alternation of greyish and reddish reticular patterns in the deep horizons indicates the hydromorphic conditions to which these soils have been submitted. On the other hand, the presence of clay-skins points to the existence of an illuviation of clay. Also slickensides are observed (Photo 3).

The extremely clayey texture of the deep horizons contrasts with the more silty aspect of the A_2 horizon whose greyish-whitish colour indicates a loss of elements. The compactation it presents is note-worth; its structure is massive and takes on a columnar pattern on outcropping (Photo 4).

The compactation of this horizon hinders the progression of erosion and gives place to a characteristic relief in which the A_2 horizon occupies the top part of this eroded landscape. The progression of erosion has caused the disappearance of this horizon in some places, and the textural horizon is laid bare at the surface where it dries and becomes very fragile. Polygonal surface cracking is frequent.

White accumulations of halloisite with a high degree of purity are frequently observed in the deeper parts of the basalts.

In the El Cepo plateau, spectacular polygonal structures up to 900 cm^2 in extension are observed on the surface (Photo 2). These are a reddish colour inside and white at the rims in relation to the disgregation pattern of the material.

A typical and representative profile of these soils is described below. The profiles "El Cepo" and "Juego de Bolas" are described in the "Guía de Campo" (Guide Book) Congreso Internacional de Suelos Volcánicos. Departamento de Edafología (1984).

Photo 1:   Erosion landscape.

Photo 2:   Polygonal weathering.

Photo 3: "Juego de Bolas" profile.

Photo 4: A$_2$ horizon (detail).

Photo 5: Ironstone.

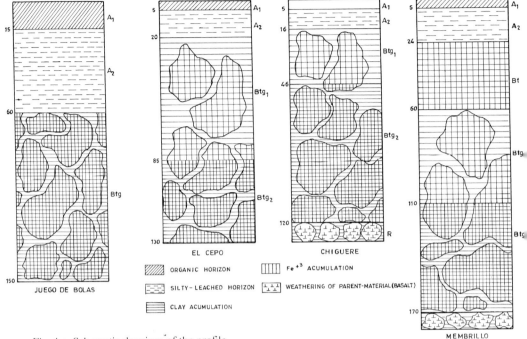

Fig. 4:   Schematic drawings of the profile.

Profile: Membrillo
Location: Plateau situated in the northern part of the island La Gomera, above the Hermigua valley
Altitude: 600 m
Distance from sea cost: 1100 m
Physiographic situation: Continuation of the El Cedro plateau (Fig. 2)
Vegetation: Eucalyptus, cypresses and pines
Erosion stage: Very advanced, with deep gulleys
Human impact: None
Parent rock: Horizontal basalts
A_1 horizon (0-5 cm):   Colour 10YR 3/3 (dark brown), clayey texture; crumb structure; abundant roots; low bulk density; positive NaF reaction.
A_2 horizon (5-24 cm):   Colour 5 YR 5/2 (greyish brown) dry; silty clay texture; massive structure with a columnar tendency, very compact; high porosity; remains of fin roots. Net transition.
Bt horizon (24-60 cm):   Colour 7,5 YR 3/4 (dark brown) dry; some yellowish faces are observed; clayey texture; polyedric structure; small bright faces with striated stress-cutans. Reddish and yellowish patches appear gradually and the background is progressively greyer. The depth of the Bt horizon varies. The sample was taken from a zone approximately 40 cm deep, although some points reach 1 m.
Btg_1 and Btg_2 horizons: These correspond to a weathered horizon where a reddish and yellowish reticular pattern with a grey background is observed, indicating drainage problems. Yellowish patches tend to predominate against a grey background in the upper part of the horizon and reddish patches in the deeper part. Clay-skins are observed; clayey texture; structure with a polyedric tendency, extremely fragile upon drying.
R horizon: Horizontal basalt with roundish exfoliation. Accumulations of halloysite are frequently found in the fissures of the basalt (Fig. 4).

The morphological characteristics of the four profiles are very similar. Profiles "Juego de

Bolas" and "El Cepo" are fully described in "Guía de Campo" International Congress on Volcanic Soils (Tenerife 1984).

Fig. 4 represents a schematic drawing of these four profiles, which can reach a maximum depth up to 10 m in the deepest one.

3. PHYSICO–CHEMICAL CHARACTERISTICS

As indicated in Table 1, these soils present a high acidity associated with high conductivity values and high percentages of Na^+ in the exchange complex.

The characteristics of the exchange complex of the A_2 horizon correspond to those of an eluvial horizon. The lowest values of exchangeable bases are found, as well as of cation exchange capacity; however, it must be pointed out that the Na^+% in the exchange complex can be very high in this horizon, as occurs in the profiles of "El Cepo" and "Chiguéré", with values above 15%, the limit established by some classification to define alkaline soils.

With regard to the total organic material (Table 2), this tends to decrease with depth. Fulvic acids predominate in the B horizon and tend to increase with depth. Humic acids, on the other hand, tend to decrease with depth.

Table 3 gives the conductivity values and soluble ions in the saturated extract. The preferential accumulation of salts in the A_2 horizon is noteworthy, with the highest value in the "El Cepo" and "Chiguéré" profiles, the most closely located from the sea shore.

The A_2 horizon frequently appears at the surface as a consequence of erosive processes; a whitish film with a higher concentration of salts is often observed in these cases.

The ions Na^+ and Cl^- predominate among the soluble ions, and no carbonates or bicarbonates are detected.

The mechanical analysis (Table 4) again shows the eluvial nature of the A_2 horizon, with a high silt content, while the clay fraction amply predominates in the Btg horizon. A good correlation is observed between the moisture retention and the granulometric composition.

Considering the advanced degree of evolution of these soils a significance can be attributed to the total analysis in the genetic process (Table 5). Indeed, high values of aluminium are observed in the deep horizons, as compared with the lower values of this element in the A_2 horizons.

The accumulation of alkaline and alkaline-earth salts observed in the total analysis of the A_2 horizon corresponds to the values observed in the saturated extract.

An important relative increase in SiO_2 and TiO_2 is also observed in the A_2 horizon. Here again are to be found the lowest values of water loss (between 105° and 1000°C). Iron, and manganese are also present in this horizon.

In general, a considerable decrease in alkaline and alkaline-earth salts along the profile, as compared with the content in these elements of the volcanic materials, is observed.

The high values of the ratio SiO_2/Al_2O_3 are due to the existence of quartz in all the profiles and do not reflect the predominance of kaolinitic clays existing in these soils.

Tab. 1: EXCHANGE COMPLEX

Profile/	pH		Exchange complex meq/100 g						%S/T	%Na/T	meq/100	
Hor.	H_2O	KCl	Ca^{++}	Mg^{++}	Na^+	K^+	S	T			Al^{+++}	H^+
JUEGO DE BOLAS												
A_1	5,6	4,9	9,8	7,5	1,3	1,3	19,9	29,1	68,4	4,5	0,1	0,
A_2	5,3	4,6	2,5	2,2	0,5	0,3	5,5	9,5	57,9	5,3	0,1	0,
Btg	4,4	3,8	2,2	2,5	1,6	0,2	6,5	20,3	32,0	7,9	2,7	0,
EL CEPO												
A_1	4,6	4,1	6,2	6,1	2,3	1,8	16,4	30,0	54,7	7,7	0,4	0,
A_2	4,3	3,9	1,8	2,4	2,7	0,3	7,2	16,2	44,4	16,7	0,4	0,
Btg_1	4,4	3,8	4,9	4,3	3,3	1,2	13,7	28,5	48,1	11,6	2,6	0,
Btg_2	4,1	3,6	2,5	2,6	2,8	0,6	8,5	30,7	27,7	9,1	6,9	1,
MEMBRILLO												
A_1	4,7	3,8	2,4	3,9	1,4	0,5	8,2	24,3	33,7	5,8	3,9	0,
A_2	4,7	4,0	1,7	1,7	1,1	0,2	4,7	16,9	27,8	6,5	2,2	0
Bt	4,6	3,7	1,9	2,7	1,0	0,1	5,7	32,2	17,7	3,1	9,2	1
Btg_1	4,4	3,7	1,3	2,3	1,1	0,1	4,8	29,4	16,3	3,7	11,1	0
Btg_2	4,3	3,5	0,7	1,4	0,8	0,1	3,0	13,5	22,2	5,9	8,0	0
CHIGUERE												
A_1	6,1	5,4	14,2	11,7	2,3	1,0	29,2	33,3	87,7	6,9	0,0	0
A_2	5,2	4,7	3,5	3,5	2,5	0,4	9,9	13,9	71,2	18,0	0,0	0
Btg_1	4,5	3,9	4,9	5,3	2,2	0,8	13,2	18,4	71,7	11,9	1,8	0
Btg_2	4,2	3,6	2,3	2,7	2,1	0,3	7,4	15,9	46,5	13,2	4,4	1
Btg_2	4,1	3,5	1,4	2,5	1,8	0,1	5,8	29,8	19,5	6,0	4,8	0

Tab. 2: ORGANIC MATTER CHARACTERISTICS

Profile/ Horizont	%C	%O.M	%N	C/N	%CAH	%CAF	AH/AF
JUEGO DE BOLAS							
A_1	4,9	8,4	0,2	24,5	1,0	1,3	0,8
A_2	1,3	2,2	0,1	13,0	0,3	0,4	0,7
Bt	0,7	1,2	0,1	7,0	0,0	0,6	0,0
EL CEPO							
A_1	2,3	3,9	0,2	11,5	0,6	0,2	3,0
A_2	1,5	2,6			0,5	0,5	1,0
Btg_1	0,4	0,7	0,1	4,0	0,4	0,7	0,6
Btg_2	1,0	1,7	0,1	10,0	0,3	0,6	0,5
MEMBRILLO							
A_1	1,2	2,1	0,1	12,0	0,2	0,6	0,3
A_2	1,0	1,7	0,1	10,0	0,0	0,7	0,0
Bt	1,1	1,9	0,1	11,0	0,4	0,6	0,7
Btg_1	0,9	1,5	0,1	9,0	0,1	0,7	0,1
Btg_2	0,6	1,0	0,0		0,1	0,9	0,1
CHIGUERE							
A_1	4,2	7,2	0,2	21,0	1,1	0,1	11,0
A_2	1,2	2,1	0,1	12,0	0,7	0,1	7,0
Btg_1	1,5	2,6	0,1	15,0	0,3	0,6	0,5
Btg_2	1,2	2,1	0,1	12,0	0,1	0,9	0,1
Btg_2	0,3	0,5	0,1	3,0	0,0	0,7	0,0

Tab. 3:   COMPOSITION OF THE SATURATED EXTRACT

Profile/ Horizont	Distance from sea	E.C. mmhos	Ca^{++}	Mg^{++}	Na$^+$	K$^+$	CL$^-$	SO$_4^-$
JUEGO DE BOLAS	1500 m							
A$_1$		2,8	0,2	0,3	0,3	0,0	1,2	0,2
A$_2$		3,9	0,2	0,3	0,6	0,0	1,1	0,1
Btg		3,3	0,4	0,4	2,0	0,0	3,0	0,0
EL CEPO	1000 m							
A$_1$		10,4	1,2	1,9	3,3	0,2	6,5	0,2
A$_2$		21,0	0,7	1,4	6,6	0,0	9,6	0,1
Btg$_1$		9,9	1,4	1,3	5,2	0,1	8,1	0,3
Btg$_2$		7,0	0,6	0,7	4,0	0,1	5,8	0,2
MEMBRILLO	1500 m							
A$_1$		1,4	0,2	0,1	0,4	0,0		0,1
A$_2$		2,3	0,2	0,2	0,6	0,0	0,8	0,0
Bt		0,7	0,1	0,1	0,2	0,0	0,3	0,1
Btg$_1$		0,9	0,1	0,1	0,2	0,0	0,5	0,1
Btg$_2$		0,7	0,2	0,0	0,2	0,0	0,2	0,2
CHIGUERE	1250 m							
A$_1$		3,8	0,5	0,7	1,1	0,1	1,7	0,2
A$_2$		9,1	0,5	0,5	1,2	0,0	3,5	0,3
Btg$_1$		4,8	0,3	0,4	2,5	0,0	3,8	0,0
Btg$_2$		3,4	0,5	0,4	2,2	0,0	2,7	0,4
Btg$_2$		3,8	0,4	0,4	2,5	0,0	3,2	0,3

Composition of the saturated extract (meq/100 g)

Tab. 4:   PHYSICAL PROPERTIES

Profile/ Horizont	Coarse sand 2000-200μ	Fine sand 200-50μ	Silt 50-2μ	Clay <2μ	pF 1/3bar	pF 15bar
JUEGO DE BOLAS						
A$_1$	1,0	4,5	29,0	62,0	36,8	23,9
A$_2$	3,4	3,1	72,0	26,5	23,4	9,2
Btg	0,6	0,5	6,0	91,0	42,4	31,6
EL CEPO						
A$_1$	0,8	5,1	19,0	74,0	35,7	25,3
A$_2$	0,1	0,1	52,0	47,0	26,4	14,3
Btg$_1$	0,5	1,4	5,0	91,0	48,1	32,8
Btg$_2$	0,3	0,7	6,0	95,0	49,9	34,6
MEMBRILLO						
A$_1$	1,8	2,3	15,5	77,0	31,1	23,3
A$_2$	1,9	2,2	50,0	46,0	28,2	15,2
Bt	0,2	0,2	5,0	97,0	40,8	30,7
Btg$_1$	0,2	0,4	6,0	98,0	45,3	30,5
Btg$_2$	0,2	0,7	6,0	94,0		
CHIGUERE						
A$_1$	1,4	7,3	35,0	54,0	42,0	23,0
A$_2$	1,6	2,3	52,5	41,0		
Btg$_1$	0,3	0,7	11,5	87,5	43,5	30,7
Btg$_2$	0,3	0,5	8,0	94,0	46,6	35,3
Btg$_2$	1,7	1,3	17,0	83,0	43,9	29,5

Mechanical analysis (%)

Tab. 5: TOTAL SOIL CHEMICAL ANALYSIS

Profile Horizont	%SiO$_2$	%Al$_2$O$_3$	%Fe$_2$O$_3$	%TiO$_2$	%K$_2$O	%Na$_2$O	%CaO	%MgO	%Mn$_3$O$_4$	Ignition loss 105/1000°	SiO$_2$/Al$_2$O$_3$
JUEGO DE BOLAS											
A$_1$	32,8	17,8	18,9	4,8	0,9	0,3	0,1	0,7	0,1	18,6	3,1
A$_2$	44,9	11,3	25,1	6,9	1,0	0,5	0,3	0,9	0,2	6,3	6,7
Btg	34,2	28,3	17,4	3,8	0,6	0,2	0,0	0,3	0,0	10,2	2,0
EL CEPO											
A$_1$	44,9	21,4	15,6	3,0	1,0	0,3	0,0	0,5	0,0	15,7	3,6
A$_2$	53,5	12,1	15,7	3,7	1,1	0,6	0,0	0,5	0,1	8,1	7,5
Btg$_1$	35,3	22,7	14,7	2,5	1,2	0,4	0,0	0,6	0,0	15,1	2,6
Btg$_2$	33,2	28,3	19,0	2,2	1,0	0,5	0,0	0,4	0,0	11,7	2,0
MEMBRILLO											
A$_1$	38,5	22,7	16,5	3,7	0,6	0,2	0,0	0,6	0,1	13,9	2,9
A$_2$	46,0	16,1	17,9	6,1	1,0	0,4	0,0	0,8	0,2	7,3	4,8
Btg	42,8	28,3	12,4	2,6	0,7	0,1	0,0	0,4	0,0	12,2	2,6
Btg$_1$	40,7	28,3	12,0	2,6	0,5	0,1	0,0	0,3	0,0	11,4	2,4
Btg$_2$	35,3	22,7	17,5	4,3	0,4	0,1	0,0	0,3	0,0	11,7	2,6
CHIGUERE											
A$_1$	36,4	17,0	17,9	5,3	0,8	0,7	0,0	0,7	0,1	17,4	3,6
A$_2$	55,6	11,3	12,9	5,9	1,0	0,6	0,0	0,5	0,1	7,0	8,3
Btg$_1$	33,1	22,7	20,9	3,8	0,9	0,3	0,0	0,5	0,0	12,7	2,5
Btg$_2$	32,9	22,7	22,3	4,3	0,6	0,4	0,0	0,3	0,0	11,4	2,4
Btg$_2$	39,6	22,7	19,5	5,2	0,5	0,2	0,0	0,3	0,1	10,7	3,0

Tab. 6: APPROXIMATE MINERALOGICAL COMPOSITION ESTIMATED BY X–RAY DIFFRACTION (PROFILES)

Profile/ Horizont	Clay type				Hematite	Goethite	Quartz	Felspar	Maghemite	Gibbsite
	1/1 Kaolinite	2/1 I	I–V	Sme.						
JUEGO DE BOLAS										
A₁	++	++	o	o	+++	+	++	++	++	++
A₂	+	++	o	o	+++	+	+++	++	+	+
Btg	++++	+	–	o	+	+	tr.	o	o	o
EL CEPO										
A₁	++	++	–	?	++++	+	++	o	o	o
A₂	+++	+++	o	o	++++	–	+++	–	o	o
Btg₁	+++	+	–	?	++++	++	+++	o	o	o
Btg₂	++	+	–	?	+	++	+	o	o	o
MEMBRILLO										
A₁	++	–	?	o	+++	+	++	o	o	?
A₂	++	–	o	o	+++	+	+++	o	o	o
Btg	+++	–	?	+	+++	+	+	o	o	o
Btg₁	+++	–	o	+	tr.	–	+	o	o	o
Btg₂	+	–	o	+	–	+++	–	o	o	o
CHIGUERE										
A₁	++	++	o	o	+++++	+	++	tr.	o	o
A₂	++	+++	o	o	+++++	++	++	–	o	o
Btg₁	–	++	o	o	+++++	++	+++	o	o	o
Btg₂	+++	+	o	–	+++++	+++	–	o	o	o
Btg₂	+	–	o	–	+	+	–	o	o	o

4. MINERALOGICAL CHARACTERISTICS

4.1. MICROSCOPIC STUDY OF THE SAND FRACTION

Within the very low percentages of sand fraction presented by these soils, opaque minerals always predominate, mainly iron and iron titaned (magnetite, ilmenite, titan magnetite, hematites and goethite) or grains that are difficult to identify and that are made up of very altered fragments of basalt and clayey-ferruginous micronodules.

Transparent minerals only appear in trace proportion: plagioclase, sanidine, quartz, feldspar and augite.

4.2. X–RAY DIFFRACTION STUDY

The mineralogy observed in these soils using X-ray diffraction techniques is indicated in Table 6, and approximate values are given. Kaolinites predominate amply among the clay minerals, having a high degree of crystallization. Halloysites are scarcely present. This in an infrequent case in the soils of the Canary Archipelago. Together with the kaolinites, some type 2/1 clay minerals are to be found, mainly illite or illite-vermiculite and, more rarely, traces of smectite are detected. Iron oxides and hidroxides are very abundant, as also hematites and goethite in particular, this last mineral increasing with depth. With regard to forms of aluminium, only gibbsite traces have been noted in some horizons. Quartz is present in all the profiles, with an important concentration in the A_2 horizons.

A separate study was carried out on the minerological composition of the oxidised and reduced zones of plinthite, no significant differences being observed except in the quartz content. In the "Juego de Bolas" profile, being very abundant in the reduced zones and insignificant in oxidised reticular patterns.

White accumulations of pure halloysite are observed in contact with the geological materials situated at greater depth which are less weathered.

4.3. STUDY WITH TRANSMISSION ELECTRON MICROSCOPE

Kaolinite is only observed, of small size, very well crystallized in perfectly hexagonal crystals, together with globular accumulations of iron oxides (probably goethite), particularly in the ironstone, and very small proportions of tubular halloysite in the form of small, short tubes.

5. MICROMORPHOLOGICAL CHARACTERISTICS

The A_1 horizon of these soils, when it appears, presents a very heterogeneous microscopic aspect. It is made up of random packing, very porous (50% approximately) of plant remains with a variable degree of preservation, charcoal fragments and small spheroidal peds (30-500 μ) in both crumbs and granules, many of which seem to have a clearly biological origin.

Coarse basic mineral components (over 12,5 μ in size) are rare and consist almost exclusively of very weathered and clayed irregular fragments of augite-plagioclase basalt with few pheno-crystals of olivine transformed into iron oxides.

The fine material is a very dark speckled brownish yellow. Many mineral crystals under 12,5 μ in size can be identified. These are mainly angular quartz crystals, usually associated as crystal new formation. Thus, their morphology and disposition lead us to think of their new pedological formation. Small crystals of magnetite, hematite, and goethite are observed as well as quartz.

The existence of clay and silty clay accumulations, grey yellow or red in colour, is one of the most important microscopic features of this horizon. They usually show a certain development of the b-fabric, and although initially these accumulations may be interpreted as papules or argilane fragments, they are most probably residual fragments of the fine material and must be considered as pedorelics because of the absence of microstratification and the similitude with the fine heterogeneous material that appears in the Btg-type deep horizons.

The A_2 horizon lacks a true fine clay fraction and forms a continuous silty and very porous unit (15-20 %) made up of many microcrystals of quartz and hematites. Thus, the microstructure may be considered not aggregated, since separate peds that may be considered as such do not exist. The porosity is made up mainly of vesicles, vughs and simple packing voids. Neither mineral nor organic coarse basic components are observed. As stated previously, no fine clay material is present. The fine material existing shows a dark grey colour at low magnifications, but with high resolution clearly shows its silty quartz hematitic nature. The quartz micrograins are between 5-10 μ in size, they are angular, frequently forming crystal new-formation and are highly birefringent in the first order of greys. The hematite microcrystals are smaller in size (1-4 μ), dark brown in colour, rounded in shape and birefringent in the second order of reds. Crystallitic b-fabric.

Pedorelicts or patches of fine material similar to those described for the A_1 horizon are also observed, as also dendritic or irregular ferruginous segregations in abundance, showing a high degree of impregnation.

All these characteristics indicate a degradation horizon in which the frequent patches of fine material are no more than relics or evidence of an intense mobilization and weathering of the fine material in hydromorphic conditions, indicating ferrolysis and redistribution of the iron sesquioxides observed (LANGOHR et al. 1983).

The Btg-type horizons situated in the deep part of these soils are characterized by a total absence of coarse basic components and are composed in their totality of a very heterogeneous fine clay fraction as regards colour and composition, typical of a deeply weathered horizon that has undergone an intense illuviation in hydromorphic conditions (FEDOROFF 1974).

These horizons are compact with variable microstructure with some zones where subangular and angular blocky peds predominate while in others the microstructure may be considered as fissured. The porosity occupying some 15% of the thin section is composed of fissures of between 50-100 μ in diameter and some irregular vughs within the peds. Neither mineral nor organic coarse basic components are observed.

The fine material, very heterogeneous in colour, is characteristic of an accumulation and degradation in hydromorphic conditions. Three different types of fine mass are identified:

a)   Intense red colour, limpid aspect, and of ferric-clay nature.

b)   Yellow in colour, limpid aspect, of clayey nature. Patches more reddish in colour are frequently found, manifesting the existence of an important mobilization of sesquioxides.

c) Speckled grey in colour and silty in nature. Some red or yellow residual patches are observed within this mass that seem to indicate in this case that, not only are the processes of mobilization more intense, but that processes of redistribution or degradation of the clays exist, giving it a silty appearance. The kevex microanalysis gives the following composition for state above each of the three different types of fine mass:

$$a: 40\% \text{ of Fe; } b: 9\% \text{ of Fe and } c: 6\% \text{ of Fe.}$$

Although in many cases the b-fabric of the fine material is undifferentiated, many zones with a birenfringence high in patches are frequently observed, typical of a sedimentation in hydromorphic conditions.

Microcrystals of quartz, hematite and magnetite are still observed as microcontrasted particles in the fine material, particularly in the silty grey zones but these do not reach the proportions of the A_2 horizon.

5.1. CHARACTERISTICS OF THE IRONSTONE

Hardened horizons that correspond to the formation of ironstone are found in the four plateaus studied. This is the first time that formations of this type have been observed in the Canary Archipelago.

The characteristics of the ironstone, the location they occupy, and their relationship to the associated soils will now be discussed.

The greater part of these hardened horizons are to be found in zones affected by intense erosive processes.

Although they occupy different positions within the plateaus they reach their greatest development in low slope zones. At the present time, as a consequence of the greater suscep-tibility to erosion of the horizons surrounding the ironstone, these are beginning to occupy a cornice position leading to "Bowal"-type situations (MAIGNIEN 1958). They are usually found directly above basaltic materials and, in some cases, a certain continuity can be observed between them.

There is variation in the extension occupied by these ironstones, as also in form and degree of hardness. They either form more or less continuous layers that appear in certain nucley of the plateaus or they are found fragmented in spots of different sizes, usually small, below 5 cm, on the surface of the soil. Only the second type has been identified in the "Juego de Bolas" plateau. The thickness and morphology of these ironstone formations vary and ranges between 0,5 and 2,5 m.

The ironstones are laminar, nodular and scoriaceous, dark grey almost black in colour, reddish, yellowish and violet, having many colours with mixtures of reds and greys, although the dark-coloured ironstones predominate. Thee are also different degrees of hardness; some are extremely hard and difficult to break up.

It is interesting to observe how in some of the less hardened laminar ironstones the sheets correspond to the red and grey veins of the Btg horizons. In some of the zones in which this type of ironstone appears at the base of a slight slope, the reddish and greyish veins in the Btg horizon are seen to present an inclination parallel to the slope.

With regard to the chemical composition, these hardened horizons present high con-tents of Fe_2O_3, the highest value being approximately 60%. The total analysis of some of these hardened horizons appears in Table 7.

Tab. 7:   TOTAL IRONSTONE CHEMICAL ANALYSIS

Sample	%SiO$_2$	%Al$_2$O$_3$	%Fe$_2$O$_3$	%TiO$_2$	%K$_2$O	%Na$_2$O	%CaO	%MgO	%Mn$_3$O$_4$	Ignition loss 105/1000°
JUEGO DE BOLAS Ironstone	13,9	14,5	59,8	2,7	0,1	0,1	–	0,1	0,0	9,5
EL CEPO Ironstone	11,8	10,4	57,2	2,5	0,1	0,2	–	0,2	0,1	13,8
MEMBRILLO Ironstone	12,8	11,2	57,6	3,1	0,1	0,2	–	0,2	0,2	12,2
CHIGUERE Ironstone	23,5	14,0	51,4	4,0	0,2	0,3	–	0,2	0,2	11,7

Tab. 8:   APPROXIMATE MINERALOGICAL COMPOSITION ESTIMATED BY X-RAY DIFFRACTION (IRONSTONE)

Sample	(Kaolinite)	Hematite	Goethite	Magnetite	Quartz	Maghemite	Gibbsite
JUEGO DE BOLAS ironstone	+	+	++++	+	0	?	0
EL CEPO ironstone	–	+	+++++	0	0	–	0
MEMBRILLO ironstone	+	?	+++++	0	++	0	tr.
CHIGUERE ironstone	+	0	+++++	0	++	0	0

In mineralogical composition (Table 8) all the ironstones have in common the predominance of goethite, being in many cases practically the only mineral identified.

In general, well crystallized kaolinites are to be found together with goethite, 2/1-type clays have not been detected.

The content of quartz and hematites varies from one ironstone to another. In some cases these minerals are not present, while in others they abound.

It is interesting to note that these hardened horizons present qualitatively the same mineralogical composition as the Btg horizons of the soils studied, with the exception of the traces of illite found in the latter.

The ironstone studied present microscopically a high porosity (15-25 %) composed mainly of irregular vughs and, to a lesser degree, vesicles and some planes.

In general, there is a total absence of coarse basic components (ø $>$ 12,5 μ) both mineral and organic.

The fine mass of the ironstone, is varied in colour but, with the exception of the intense ferruginization affecting it, the background colour is yellowish red with a slightly speckled aspect due to the presence of microcontrasted particles consisting of microcrystals (ø $<$ 12,5 μ) of quartz, hematites, magnetite and micronodules of iron oxides (probably goethite). Although in some ironstone the fine mass is completely isotropic due to the masking of the birefringent domains by ferruginous segregations, in others, where the ferruginization has been less intense, the b-fabric is very developed, with elongated domains, very birefringent in the second order reds and yellows.

However, the microscopic features most characteristic of the ironstone are those related to the ferruginous segregations that impregnate, often to a high degree, the entire clayey mass. These segregations are dark red in colour and varied in shape: irregular, dendritic, elongated, etc. and in some case the ferruginization is seen to begin to isolated oval nodules from the yellowish clayey mass that are still present after the ferruginization has impregnated the entire mass.

5.2. CLASSIFICATION

The Soil Taxonomy, fails for the classification of this kind of paleosols. Therefore we will confine ourselves to identifying the diagnostic horizons.

The four profiles have an "ochric epipedon". The subsuperficial horizons have the properties of the argillic horizon. The thin section shows enough oriented clay as to define the argillic horizon, however, but in terms of percentage is relatively low. Such a high residual clay content in the B makes difficult the observation of the illuviation cutans, especially in the presence of slickenside.

6. DISCUSSION

It is, indeed, evident that horizontal or subrecent basalts that constitute the parent material of these soils have undergone intense weathering in an originally well-drained medium and in conditions of higher pluviometry than at the present time.

Weathering of plagioclase-augitic basalts in the above-mentioned conditions in Gomera, has led to the new formation of clay minerals of the network 1:1 (Kaolinite), and to the individualization of crystalline sesquioxides of iron, titanium and manganese. The total elimination of bases and the nature of the secondary minerals formed indicate clearly that the process is one of ferrallitic weathering (TEJEDOR et al. 1978, CHATELIN 1974).

This practically total weathering of the parent material produces an intense clay formation with the consequent change in the drainage conditions of the medium. This weathering is very progressive, and the gradual transition between the basalt and the soil can readily be observed in the field.

As a consequence of the formation of an impermeable clay horizon, the rainwater cannot drain readily and tends to accumulate, forming a perched water, that fluctuates in the soil, according to the seasons. At this stage, the processes taking place are those of hydromorphy and clay illuviation. In other words, conditions appear that enhance the phenomena of oxidation and reduction, where as in the first phase only oxidative phenomena were acting.

A state is thus reached in which the processes of reorganization and movement of clays and iron attain their maximum expression given rise to the thick Btg horizons observed.

The above processes lead to the formation of a A_2 horizon. The decrease of clay in the A_2 in relation to A_1 is probably one to two related processes: ferrolysis and lateral movement on the top of the B horizon. When moisture cannot infiltrate rapidly enough into the B, the temporary perched water table formed, promotes ferrolysis and at the same time removes the weathering product by lateral how. The ferrolysis process is responsible for the low pH values observed. The fact that free Na^+ soluble salt are always formed in the soil solution, hinders the dissocation of Na^+ from the complex; a process which favours the maintenance of the low pH values through the effect of the common ion (DAN et al. 1968).

Another important characteristic of the A_2 horizon is the presence of great amounts of quartz in the loam fraction. The presence of such a high amount of quartz in a basalt derived soil is difficult to interpret.

Three arguments can be used to explain the presence and accumulation of quartz in this material: eolian or neoformation origin and a combination of both.

The Sahara dust coming to the islands regularly from the desert can explain the eolian origin, and the age of these old surface may account for its high accumulation.

However, neoformation is also a possible explanation, but to confirm this second source of quartz, it must by performed an oxigen isotopic ratios study of this material.

The formance and accumulation of quartz in volcanic materials through a combination of those two processes, have been suggested by different authors elsewhere (YAALON et al. 1973, MORRAS 1983).

At the present time, samples of this quartz, are in the process of being analyzed by the technique of CLAYTON & MAYEDA (1963) for oxigen isotopic composition of quartz.

A change in the climatic conditions that took place during the Quaternary toward, a lesser pluviometry, has been the cause of another fundamental process: the salinization and sodification of the exchange complex observed in all of these profiles.

The fact that the salts are invariably sodic, leads us to think they are of marine origin. Besides the fact that the profiles more closed to the sea shore are the more saline confirm this idea (Table 3). This phenomena of airborne salt accumulation, and the influence of the distance from sea has been studied elsewhere (YAALON 1964, YAALON & LOMAS 1970, etc.).

The preferential accumulation of these salts in the A_2 horizon, due to a textural discontinuity, explains the columnar structure of the horizon.

During this entire process, the intense movement and reorganization of iron sesquioxides and their ageing and recrystalization, are the processes responsible for the formation of plinthites and even of goethitic ironstone, as an extence case of accumulation of iron, either upon the weathered mass itself or over the parent material, in those zones of the landscape most favourable to the oxidation and precipitation of iron.

Again, changes towards aridization of the climate accelerated even more the processes of oxidation and ageing of the iron compound and therefore the formation of plynthite and ironstone.

All the above characteristics of these soils confirm the existence of superimposed genetic processes, among which the most notable is a recent process of salinization accompanied by the existence of sodium in the exchange complex.

The superposition of a process of salinization upon a ferrallitization is a phenomenon of great pedological interest and prompts us to consider a recent climatic change together with the proximity of the sea.

Polygenetic process of this nature have been described in the literature elsewhere. There is a rich Australian literature on old weathering profils, now in arid or semiarid climates (CHURCHWARD & GUNN 1983).

The characteristics of these soils allow us to establish that the following succession of processes intervened in their genesis:

ILLUVIATION

FERRALLITIZATION   HYDROMORPHY   SALINIZATION

PLINTHIZATION

$\longrightarrow$

Time

Fig. 5:   Sequence of processes

BIBLIOGRAPHY

ALEIXANDRE, T. & PINILLA, A. (1968): Algunas modificaciones en las técnicas aplicadas al estudio mineralógico de las fracciones gruesas o arenas. Anales de Edaf. y Agrob., **27**, 564-567.
BRAVO, T. (1964): Estudio geológico y petrográfico de la isla de La Gomera. Estudios Geológicos, **XX**, 1-56.
BOWER, C.A., RETTEMEIER, R.F. & FIREMAN, M. (1952): Exchangeable cation analysis of saline and alkaline soils. Soil Science, **73**, 251-261.
BREWER, R. (1976): Fabric and mineral analysis of soils. Robert E. Keieger Publishing Comp., Hunsington, N.Y., 482 p.
CENDRERO, A. (1981): Estudio geológico y petrológico del complejo basal de la isla de La Gomera (Canarias). Estudios Geológicos, **XXVII**, 3-73.
CHATELIN, Y. (1974): Les sols ferrallitiques. (III) L'alteration. ORSTOM, Nr. **24**, 144 p.
CHURCHWARD & GUNN (1983): Stripping of deep weathered mantles and its significance to soil patterns. In: Soils – an Australian View point. CSIRO, Melb. and Acad. Press, London, 73-81.
CLAYTON, R.N. & MAYEDA, T.K. (1963): The use of bromine penta-fluoride in the extraction of oxygen from oxides and silicates for isotopic analyses. Geochim. Cosmochim. Acta, **27**, 43-52.

DAN, J., YAALON, D.H. & KOYUMDJISKY, H. (1968): Catenary soil relationship in Israel. A. the Netanya catena on coastal dunes of the Sharon. Geoderma, **2**, 95-120.

DEPARTAMENTO DE EDAFOLOGIA (1984): Guía de campo. Congreso Internacional de suelos volcánicos. Universidad de La Laguna, Tenerife, julio 1984.

FEDOROFF, N. (1974): Interaction between hydromorphy and lessivage. – Example of a sequence of soils lessives – with increasing hydromorphy over Quaternary silt in the south Paris Bassin. Pseudogley and Gley, 295-305.

KONONOVA, M.M. (1966): Soil organic matter. Pergamon Press, Oxford.

LANGOHR, R. & PAJARES, G. (1983): The chronosequence of pedogenic processes in Fraglossudalfs of the Belgian loess belt. In: Soil Micromorphology (P. Bullock and C.P. Murphy Ed.) vol. **2**, 503-510.

MAIGNIEN, R. (1958): Le cuirassement des sols en Guinée. Mem. Carte Geol. Alsace–Lorraine. Nr. 16.

MINISTERIO DE AGRICULTURA (1974): Métodos oficiales de análisis de suelos y aguas. Madrid.

MORRAS, H. (1983): Some properties of degraded argillans from A_2 horizons of solodic planosols. In: Soil micromorphology (P. Bullock and C.P. Murphy Ed.) vol. **2**, 575-581.

PEREZ MATEOS, J. (1965): Análisis mineralógico de arenas. Manuales de Ciencia Actual Nr. **1**, C.S.I.C. Madrid.

QUANTIN, P., TEJEDOR, M.L. & FERNANDEZ CALDAS, E. (1978): Observations sur la presence de sols ferrallitiques derivés de materiaux volcaniques aux Iles Canaries. Cah. ORSTOM, ser. Pedol. **4**, 75-84.

RICHARDS, L.A. (1954): Diagnosis and improvement of saline and alkali soils. Agricultural Handbook, Nr. **60**, p. 100. USDA.

SZABOLCS, J. (1979): Review of research on salt-affected soils.

U.S.D.A. (1975): Soil Taxonomy. A basic system of soil classification for making and interpretation soil surveys. 175 p.

WALKLEY, A. & BLACK, I.A. (1933): An examination of a rapid method for determining organic carbon and nitrogen in soils. Jour. Agric. Sci., **25**, 598-609.

YAALON, D.H. (1964): Airborne salts as an active agent in pedogenetic processes. 8th International Congress of soil science, Bucharest, Romania, 1964. Vol. **V**, 997-1000.

YAALON, D.H. & LOMAS, J. (1970): Factors controlling the supply and the chemical composition of aerosols in a near-shore and coastal environment. Agricultural Meteorology – Elsevier Publishing Company, Amsterdam, 443-454.

YAALON, D.H. & GANOR, E. (1973): The influence of dust on soils during the Quaternary. Soil Science, Vol. **116**, Nr. 3, 146-155.

Address of authors:
M.L. Tejedor Salguero, C. Jimenez Mendoza, A. Rodriguez Ridriguez and E. Fernandez Caldas, Departamento de Edafologia, Universidad de La Laguna, Tenerife, Canary Islands, Spain.

CATENA SUPPLEMENTS

CATENA SUPPLEMENT 1 (1982): **Dan H. Yaalon (Ed.):**
Aridic Soils and Geomorphic Processes
Selected Papers of the International Conference of the International Society of Soil Science, Jerusalem, Israel, March 29 - April 4, 1981
ISSN 0722–0723 / ISBN 3–923381–00–X
Price DM 95,- / US$ 55,-

CATENA SUPPLEMENT 2 (1982): **F. Ahnert, H. Rohdenburg & A. Semmel:**
Beiträge zur Geomorphologie der Tropen (Ostafrika, Brasilien, Zentral– und Westafrika) – Contributions to Tropical Geomorphology
ISSN 0722–0723 / ISBN 3–923381–01–8
Price DM 120,- / US $ 69,-

CATENA SUPPLEMENT 3 (1983): **H.R. Bork & W. Ricken:**
Bodenerosion, Holozäne und Pleistozäne Bodenentwicklung – Soil Erosion, Holocene and Pleistocene Soil Development
ISSN 0722–0723 / ISBN 3–923381–02–6
Price DM 105,- / US $ 60,-

CATENA SUPPLEMENT 4 (1983): **Jan de Ploey (Ed.):**
Rainfall Simulation, Runoff and Soil Erosion
ISSN 0722–0723 / ISBN 3–923381–03–4
Price DM 120,- / US $ 75,-

CATENA SUPPLEMENT 5 (1984): **Asher P. Schick (Ed.):**
Channel Processes – Water, Sediment, Catchment Controls
ISSN 0722–0723 / ISBN 3–923381–04–2
Price DM 110,- / US $ 60,-

CATENA SUPPLEMENT 6 (1985): **Peter D. Jungerius (Ed.):**
Soils and Geomorphology
ISSN 0722–0723 / ISBN 3–923381–05–0
Price DM 120,- / US $ 60,-

DAN H. YAALON (ED.)

ARIDIC SOILS and GEOMORPHIC PROCESSES

SELECTED PAPERS of the INTERNATIONAL CONFERENCE
of the INTERNATIONAL SOCIETY of SOIL SCIENCE
Jerusalem, Israel, March 29 – April 4, 1981

CATENA SUPPLEMENT 1, 1982

Price: DM 95,–/US $ 55.–
Special rate for subscription until June 30, 1982: DM 66,50/US $ 38.–
(available from the publisher only)
ISSN 0722–0723 / ISBN 3–923381–00–X

This CATENA SUPPLEMENT comprises 12 selected papers presented at the International Conference on Aridic Soils – Properties, Genesis and Management – held at Kiryat Anavim near Jerusalem, March 29 – April 4, 1981. The conference was sponsored by the Israel Society of Soil Science within the framework of activities of the International Society of Soil Science. Abstracts of papers and posters, and a tour guidebook which provides a review of the arid landscapes in Israel and a detailed record of its soil characteristics and properties (DAN et al. 1981) were published. Some 49 invited and contributed papers and 23 posters covering a wide range of subjects were presented at the conference sessions, followed by seven days of field excursions.

The present collection of 12 papers ranges from introductory general reviews to a number of detailed, process oriented, regional and local studies, related to the distribution of aridic soils and duricrusts in landscapes of three continents. It is followed by three papers on modelling and laboratory studies of geomorphic processes significant in aridic landscapes. It is rounded up by a methodological study of landform–vegetation relationships and a regional study of desertification. Additional papers, related to soil genesis in aridic regions, are being published in a special issue of the journal GEODERMA.

D.H. Yaalon
Editor

F. Ahnert, H. Rohdenburg & A. Semmel:

BEITRÄGE ZUR GEOMORPHOLOGIE DER TROPEN (OSTAFRIKA, BRASILIEN, ZENTRAL- UND WESTAFRIKA) CONTRIBUTIONS TO TROPICAL GEOMORPHOLOGY

CATENA SUPPLEMENT 2, 1982
Price: DM 120,–/US $ 69.–
ISSN 0722–0723 / ISBN 3–923381–01–8

JAN DE PLOEY (ED.)

RAINFALL SIMULATION
RUNOFF AND
SOIL EROSION

CATENA SUPPLEMENT 4

Jan de Ploey (Ed.)

RAINFALL SIMULATION, RUNOFF and SOIL EROSION

CATENA SUPPLEMENT 4, 1983

Price: DM 120,– / US $ 75.
Special rate for subscription until July 31, 1983 (available from the publisher only)
DM 87,50 / US $ 52.50

ISSN 0722–0723 ISBN 3–923381–03–4

...is CATENA–Supplement may be an illustration of present-day efforts made by ...rphologists to promote soil erosion studies by refined methods and new conceptual :hes. On one side it is clear that we still need much more information about erosion sys-...hich are characteristic for specific geographical areas and ecological units. With respect ...bjective the reader will find in this volume an important contribution to the knowledge ...e soil erosion, especially in typical sites in the Mediterranean belt, where soil ...tion is very acute. On the other hand a set of papers is presented which enlighten the ...nt role of laboratory research in the fundamental parametric investigation of pro-...i.e. erosion by rain. This is in line with the progressing integration of field and ...ory studies, which is stimulated by more frequent feed-back operations. Finally we ...draw attention to the work of a restricted number of authors who are engaged in the dif-...aboration of pure theoretical models which may pollinate empirical research, by ...ng new concepts to be tested. Therefore, the fairly extensive publication of two papers ...LLING on soil creep mechanisms, whereby the basic force-resistance problem of ...is discussed at the level of the individual particles.

...the other contributions are focused mainly on the processes of erosion by rain. The ...ainfall simulators is very common nowadays. But investigators are not always able to ...e full fall velocity of waterdrops. EPEMA & RIEZEBOS give complementary infor-...on the erosivity of simulators with restricted fall heights. MOEYERSONS discusses ...erosion under oblique rain, produced with his newly-built S.T.O.R.M–1 simulator ...portant contribution may stimulate further investigations on the nearly unknown ...of oblique rain. BRYAN & DE PLOEY examined the comparability of erodibility ...ements in two laboratories with different experimental set-ups. They obtained a ...:ross ranking of Canadian and Belgian topsoils.

Both saturation overland flow and subsurface flow are important runoff sources under the rainforests of northeastern Queensland. Interesting, there, is the correlation between soil colour and hydraulic conductivity observed by BONELL, GILMOUR & CASSELLS. Runoff generation was also a main topic of IMESON's research in northern Morocco, stressing the mechanisms of surface crusting on clayish topsoils.

For southeastern Spain THORNES & GILMAN discuss the applicability of erosion models based on fairly simple equations of the "Musgrave-type". After Richter (Germany) and Vogt (France) it is TROPEANO who completes the image of erosion hazards in European vineyards. He shows that denudation is at the minimum in old vineyards, cultivated with manual tools only. Also in Italy VAN ASCH collected important data about splash erosion and rainwash on Calabrian soils. He points out a fundamental distinction between transport-limited and detachment-limited erosion rates on cultivated fields and fallow land. For a representative first order catchment in Central–Java VAN DER LINDEN comments contrasting denudation rates derived from erosion plot data and river load measurements. Here too, on some slopes, detachment-limited erosion seems to occur

The effects of oblique rain, time-dependent phenomena such as crusting and runoff generation, detachment-limited and transport-limited erosion including colluvial deposition, are all aspects of single rainstorms and short rainy periods for which particular, predictive models have to be built. Moreover, it is argued that flume experiments may be an economic way to establish gross erodibility classifications. The present volume may give an impetus to further investigations and to the evaluation of the proposed conclusions and suggestions.

Jan de Ploey

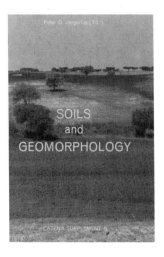

Peter D. Jungerius (Ed.):

Soils and Geomorphology

CATENA SUPPLEMENT 6 (1985)

Price DM 120,- / US $ 60,-

ISSN 0722-0723 / ISBN 3-923381-05-0

It was 12 years ago that CATENA's first issue was published with its ambitious subtitle "Interdisciplinary Journal of Geomorphology – Hydrology – Pedology". Out of the nearly one hundred papers that have been published in the regular issues since then, one-third have been concerned with subjects of a combined geomorphological and pedological nature. Last year it was decided to devote SUPPLEMENT 6 to the integration of these two disciplines. Apart from assembling a number of papers which are representative of the integrated approach, I have taken the opportunity to evaluate the character of the integration in an introductory paper. I have not attempted to cover the whole bibliography on the subject: an on-line consultation of the Georef files carried out on 29th October, 1984, produced 3627 titles under the combined keywords 'geomorphology' and 'soils'. Rather, I have made use of the ample material published in CATENA to emphasize certain points.

In spite of the fact that land forms as well as soils are largely formed by the same environmental factors, geomorphology and pedology have different roots and have developed along different lines. Papers which truly emanate the two lines of thinking are therefore relatively rare. This is regrettable because grafting the methodology of the one discipline onto research topics of the other often adds a new dimension to the framework in which the research is carried out. It is the aim of this SUPPLEMENT to stimulate the cross-fertilization of the two disciplines.

The papers are grouped into 5 categories: 1) the response of soil to erosion processes, 2) soils and slope development, 3) soils and land forms, 4) the age of soils and land forms, and 5) weathering (including karst).

P.D. Jungerius